THE
SCIENCE
OF SUPERCARS

THE
SCIENCE
OF SUPERCARS

Introduction by David Coulthard

**INCLUDING INTERVIEWS
WITH THE BIGGEST NAMES
IN SUPERCAR DESIGN**

THE TECHNOLOGY THAT POWERS
THE GREATEST CARS IN THE WORLD

Martin Roach, Neil Waterman and John Morrison

FIREFLY BOOKS

A FIREFLY BOOK

Published by Firefly Books Ltd. 2018

First printing

Publisher Cataloging-in-Publication Data (U.S.)

Library of Congress Control Number: 2018934535

Library and Archives Canada Cataloguing in Publication

Roach, Martin, author
    The science of supercars : the technology that powers the greatest cars in the world / Martin Roach, Neil Waterman & John Morrison.
    Includes index.
    ISBN 978-0-228-10090-4 (hardcover)
    1. Sports cars. I. Waterman, Neil, author II. Morrison, John (Automobile racing driver), author III. Title.
TL236.R54 2018   629.222'1   C2018-901143-2

Published in the United States by
Firefly Books (U.S.) Inc.
P.O. Box 1338, Ellicott Station
Buffalo, New York 14205

Published in Canada by
Firefly Books Ltd.
50 Staples Avenue, Unit 1
Richmond Hill, Ontario L4B 0A7

Printed and bound in China

First published by Mitchell Beazley,
a division of Octopus Publishing Group Ltd
Carmelite House
50 Victoria Embankment
London EC4Y 0DZ

Martin Roach, Neil Waterman and John Morrison have asserted their rights under the Copyright, Designs and Patents Act 1988 to be identified as the authors of this work.

Commissioning Editor: Joe Cottington
Senior Editor: Alex Stetter
Designer: Jeremy Tilston
Junior Designer: Jack Storey
Senior Production Controller: Allison Gonsalves

Front cover image © Koenigsegg Automotive AB
Back cover image © Bugatti Automobiles S.A.S.

CONTENTS

AUTHORS' NOTE

This book is not an encyclopedic history of every single supercar ever made. Nor is it a book about all of the science and engineering behind every innovation on a motor car. Its ambition is to highlight the key elements of supercar science that have shaped the way these amazing vehicles are made, including those that have also made a significant impact on the mass-market automotive world. The major milestones in supercar science are told through the firsthand recollections of some of the genre's most iconic figures, and all interviews included herein are exclusive to the authors.

Not every piece of modern-day car science has evolved through supercars, but there are dozens of examples within these pages that illustrate just how important and influential supercars have been, and no doubt will continue to be.

LEFT_The Bugatti Chiron has pushed the science of supercars to the boundaries of physics.

INTRODUCTION BY DAVID COULTHARD

The very best supercars combine engineering, science and art. There is the individual look and style of a certain marque, such as a Ferrari, Bugatti or Mercedes, but making a new car instantly recognizable isn't enough. These manufacturers have to marry the art of styling with absolute engineering technology, plus the science of great performance. When they get that right, these supercars basically represent the equivalent of high performance aeronautics. These machines are the NASA of the automotive industry.

The science hasn't always been so comprehensively advanced, as this book will illustrate. Early technology was rather protracted and at times slow to develop. In addition, many of the super-sports cars of yesteryear were developed with road and track credentials. Obviously, as motorsport has advanced to new pinnacles, those cars no longer can apply to the road; however, supercars have also advanced independently of the race track and are now capable of staggering performance. So the challenge for the designers and engineers behind these great machines is finding a way to maximize the technology and scientific understanding that is out there – on subjects such as power-to-weight ratios, centres of gravity, aerodynamics, thermodynamics and so on – but with a real-world use-ability. It's not just the engineering science either – the advances in materials such as alloys and composites all play an absolutely crucial part in this evolution. These are halo cars for the manufacturers, but the benefits of the science behind them out in the wider automotive world are manifold.

I think we are at a very rich place today in terms of high-performance sports cars, as there exists an incredibly vibrant supercar and hypercar market. These are iconic pieces of engineering that look incredible, sound amazing and have stunning performance – how that is achieved through cutting-edge science and ingenuity is the focus of this book. The very best examples of these supercars go beyond anything that has been seen before. These are remarkable science projects that will also be useable road cars and that in itself is all credit to human ingenuity.

RIGHT_ The construction of a modern supercar requires a level of precision more akin to a laboratory than a factory – as witnessed here at Bugatti in Molsheim, France.

CHAPTER 1

THE BATTLE FOR POWER

The formative decades of the science of the car – and by later association, the supercar – are all about power and invention.

OPPOSITE_Karl Benz's three-wheeled Patent Motorwagen of 1886 is the genesis of the car and, by definition, the supercar. Many of the principles behind the science of this pioneering vehicle shaped how supercars would be made in the future.

In the years leading up to the introduction of the motor car in 1885, transport was essentially horse-drawn or steam-powered. Way back in 1769, French inventor Nicolas-Joseph Cugnot had created a steam-powered, self-propelled vehicle – regarded by some historians as essentially an automobile – but over the next 150 years that initial technology proved cumbersome, not easily scaled down for personal use and potentially explosive. This was partly because the science behind these engines had been designed almost exclusively for stationary factory use, to power the exponential expansion of the Industrial Revolution and beyond. However, as the closing years of the 19th century approached, inventors, scientists and engineers began to toy with finding a mobile, light and safe alternative that could be used on a personal basis – not by 1,500 passengers on a steamship, or 400 ticket holders on a train, or even 80 people on a tram, but by one person on their own. As steam sailed off to become the choice of power for rail and sea, the finest mechanical minds began to search intensely for the ideal method of propulsion for the individual.

RIGHT_Mercer Raceabout, 1912

Over the next decade or so, an immense and at times manic battle played out between engineers and inventors in hundreds of relatively primitive workshops, factory anterooms and even privately owned sheds and store-rooms across the world. The European mainland was at the forefront of the innovation. The history books tell us that the man who got there first was Karl Benz, with his much-documented, three-wheeled, tiller-steered Motorwagen, built in 1885 and patented the year after, which was driven by a single cylinder internal combustion engine that generated just ⅔bhp. However, this achievement did not mean that petrol and internal combustion could instantly monopolize the so-called "horseless carriage". Many alternative power sources jostled fiercely for dominance: traditional coal-fired engines proved far too heavy; gas-powered engines were too dangerous and volatile for the small scale required; bizarrely, a few engineers even experimented with clockwork propulsion. However, despite Benz's initial triumph, there quickly emerged one clear favourite to be the most likely winner: electric power.

The first crude electric motor had been built by a Slovakian-Hungarian priest back in 1827 and, within a few years, Scotsman Robert Anderson had invented a battery-powered electric carriage. Sixty years later, this fledgling science had progressed sufficiently for London, New York and Paris to all boast fleets of battery-powered taxis. A major player in the future world of supercars, Ferdinand Porsche, created the Lohner-Porsche Mixte Hybrid in 1898, which caused a sensation.

"Benz's Motorwagen generated just ⅔bhp."

As America entered the battle and pioneered mass-market automotive production, companies such as Studebaker offered electric vehicles in their ranges. However, the primitive batteries in these early cars were ultimately very large and cumbersome – the same problem faced by electric cars over 100 years later. Also, many early batteries were not rechargeable, which introduced cost and inconvenience. The lack of an electricity network across the modernized world was also a huge stumbling block. So, although electric power initially seemed to be the way forward, it would ultimately fall by the wayside.

Steam power actually survived in cars for another two decades, and was refined to become stable and miniaturized – a good steam engine was simpler than an internal combustion equivalent, with far fewer moving parts. However, like the fledgling electric cars, steam vehicles suffered chronic limitations of range – a word that would come back to haunt car manufacturers more than a century later. Steam cars also took up to 45 minutes to get started. Although early petrol cars were not exactly "get in and go", they were still much quicker.

"Steam vehicles suffered chronic limitations of range."

RIGHT_Thomas Edison gets a lift from his son, Thomas Edison Jr, in a 1903 electric Studebaker.

Increasingly then, the inventors, designers and engineers turned towards the so-called "internal combustion engine" as their preferred method of power. Like the car itself, there were many rivals to lay claim to creating the first internal combustion engine, but it is generally accepted that the compressed charged, four-cylinder engine patented by Nikolaus Otto in 1876, working with the German engineers Gottlieb Daimler and Wilhelm Maybach, is the most relevant moment. Otto later parted company with Daimler and Maybach amid considerable disagreement, but the latter duo then created a high speed petrol-fuelled engine and also the first four-wheeled car (the Motorwagen had only three wheels).

Early internal combustion engine sales had been predominantly for boats, but the potential for personalized transport was clear. As the weight, range and charging issues perennially diluted the electric motor's potential, the internal combustion engine would finally win the battle as the car's power unit of choice. It was a victory that would see this method used in motor-vehicles largely unchallenged for more than a century. Even in the modern world of space exploration, air travel, computers, the internet et al, the huge majority of motor-vehicles on the planet – including until recently supercars – are still using a science that was first created while Queen Victoria was on the throne in England.

"The internal combustion engine became the car's power unit of choice."

To investigate this triumph, the first and obvious question to ask is this: why was the internal combustion engine the victor in the battle for power? Perhaps the secret to the internal combustion engine's popularity is in the name – unlike the steam engine (external combustion), the fuel is burned *internally*.

A key initial challenge was finding a fuel source that was reliable and safe. Gunpowder had been tried, as well as a mix of moss, coal-dust and resin, plus hydrogen and coal gas, but engineers settled on petrol because it was far more stable. There was also greater access to petrol – filling stations did not yet exist, but gasoline was available at pharmacies and industrial chemical suppliers, who stocked small quantities as a cleaning agent. Future supply was also not an issue as the first oil wells had been drilled in the 1850s and the underground reserves appeared to be potentially huge. The petrol-powered, internal-combustion-engined car was up and running.

How do these early history lessons relate to the science of supercars? Drilling further down into the early science and history of the car, you can begin to spot the genesis of the supercar. Progress continued apace – numerous companies vied for the title of the first car manufacturer and there was sufficient interest in the industry that the magazine *Autocar* was founded in 1895. In the same year, Britain's first motor vehicle exhibition took place in Tunbridge Wells, despite the fact there

THE INTERNAL COMBUSTION ENGINE

An internal combustion engine is so called because the explosions that generate the power happen inside the engine. There are four key actions of a four-stroke internal combustion engine: Suck, Squeeze, Bang, Blow. Typically, these four actions will be performed for every one revolution of the engine. But what exactly is happening inside each cylinder during these four actions?

_001 Suck – The piston is drawing in an air and fuel mix by moving down the inside of the cylinder and creating a vacuum. This is then filled by the air and fuel mix, which arrives through an inlet valve.

_002 Squeeze – At the bottom of the stroke of the piston inside the cylinder, the inlet valve closes, so that when the piston moves back up inside the cylinder, the air and fuel mix is squashed. By compressing the air and fuel mix in this way, it becomes more volatile.

_003 Bang – Right at the top of the stroke inside the cylinder, a spark plug ignites the compressed air and fuel mix, causing an explosion. This explosion immediately propels the piston back down the cylinder.

_004 Blow – On the return stroke back up the cylinder, the exhaust valve opens and burned fuel gases are released. This cylinder is now empty of fuel and the cycle can begin to repeat itself.

All supercar internal combustion engines follow this pattern.

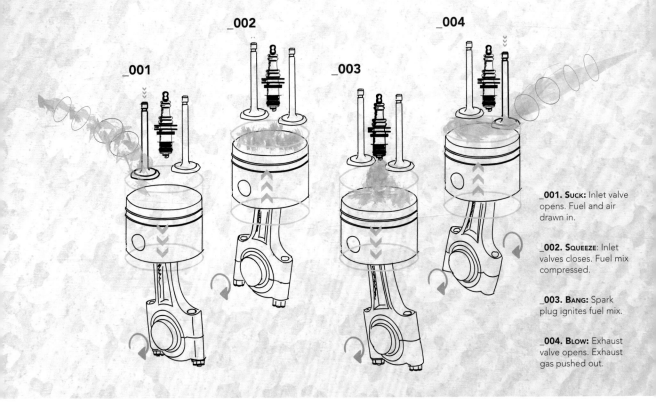

_001

_002

_003

_004

_001. Suck: Inlet valve opens. Fuel and air drawn in.

_002. Squeeze: Inlet valves closes. Fuel mix compressed.

_003. Bang: Spark plug ignites fuel mix.

_004. Blow: Exhaust valve opens. Exhaust gas pushed out.

were only 15 cars registered on British roads. At this stage, potential customers were most concerned about the reliability and efficiency of this new technology, so the early years of the motor car were more about endurance, with races such as the first London to Brighton Run by the British Motor Car Club and the Paris–Toulouse–Paris race.

However, it is the issuing of the first speeding ticket – two claims exist, one at 6mph and one at 8mph – that perhaps pre-empts the emergence of the supercar and also reveals the motivation behind their science: *a thirst for speed*. This is the crux of the fast car in these early days – as soon as there was the science to move individuals in small automobiles, certain more risk-taking members of the public started to think about going fast. Instantly synonymous with this notion was the idea of going *faster* than someone else.

In 1865 the Highways and Locomotives Act had set the speed limit in the UK at 4mph in the country and 2mph in towns, but required a "man with a red flag" to walk 60 yards in front of

BELOW_With its striking shape, the 1910 Sunbeam Nautilus represents a very early, if rather rudimentary, recognition of the fact that aerodynamics was important to performance.

the vehicle. The repealing of this law in 1896 upped the limit to 14mph, which was, in some sense, a symbolic admission that the technology behind these cars was now – metaphorically and literally – unstoppable, but also that the public's desire for faster speeds was similarly relentless. The race to build the fastest car was now well and truly on …

The urge to be the fastest – rather than just fast – would underpin supercar science for the next 100 years or more. The year 1898 saw the very first world land speed record, when Frenchman Gaston de Chasseloup-Laubat reached 39.24mph in a Jeantaud electric car. The following year, Chasseloup-Laubat's great rival, the Belgian Camille "The Red Devil" Jenatzy drove a torpedo-shaped, tiller-steered electric vehicle by the name of "La Jamais Contente" at 105kmh, the very first car to exceed the 100kmh mark. Within a decade the speed record would stand at over 127mph; by 1927 it would be over 200mph.

"In 1865, the speed limit was 4mph in the country and 2mph in towns."

BELOW_By the Roaring Twenties, the general public had become increasingly fascinated with the speed and the glamour of motorsport.

There was also a dark early reminder that high speed can have dire consequences. In 1898 the first fatality of a car driver was recorded when Henry Lindfield of Brighton died from shock the day after losing his leg when his car overturned. *Autocar* attributed the death to "excessive speed". Even so, speedometers were not even an optional extra on cars until a couple of years later.

The first years of the 20th century were a constant sequence of inaugural moments in the history of speed: in 1902 the (later Royal) Automobile Club of Great Britain organized the first race on British soil at Bexhill-on-Sea, attracting hundreds of spectators; the first Brighton Speed Trial took place in 1905, with the winner driving a 90bhp Napier; as it became apparent that city and public road racing was inherently very dangerous for all concerned, bespoke circuits sprang up, such as the very first at Brooklands in Surrey, with two huge banked stretches of a concrete circuit. Therein the concept of motorsport had been

> "The first years of the 20th century were a constant sequence of inaugural moments in the history of speed."

ignited and that would be inextricably linked to the science of supercars for the majority of the next 100 years.

So before the tragic devastation of World War I, there were fast cars, motor races, racing circuits and rival manufacturers creating ever quicker vehicles. Companies such as Panhard, Darracq and Brouhot were important early innovators in terms of sheer power and speed. But were there any supercars? And if so, what was the science behind them? Arguably the 35bhp Mercedes of 1901 might lay claim to being the first genuine "supercar". In December of the previous year, German-born businessman and motor car agent Emil Jellinek asked Daimler to produce a car that was lower and more powerful than the norm, so he could compete in hill climbs (a very popular early race format). The resulting 35bhp car that was launched in 1901, named after Jellinek's daughter, Mercedes, was the very first vehicle to bear that label and proved an instant pedigree race winner. So convinced was Jellinek of the commercial potential for a super-fast car that he placed an order on the spot for 36 vehicles, albeit with a very short deadline of just eight months. Despite the compressed time-frame to design and build a brand-new car, Daimler delivered what many automotive historians consider to be the first true supercar. The science behind the Mercedes would support that: a low centre of gravity from its low-slung, pressed-steel chassis, a low-mounted, high performance and lightweight engine, an aluminium crankcase, mechanical inlet valves (rather than the more common atmospheric type), improved drum brakes, two carburettors and even a honeycomb radiator with 8,000 tubes inside. The car was the first of a generation of faster cars that were evolving away from just being a "horse and carriage" style design. A later derivative 60hp car was also ahead of its time and could reach an almost unheard of top speed of 60mph.

There are other cars that could be argued to have pushed the science so significantly that they might be considered the very earliest of the supercar breed: the Daimler 80 PS of 1910 was designed by Dr Porsche, could reach 88mph and boasted bodywork with early signs of streamlining. Its output was considered so powerful that chains were used in favour of shafts (which designers feared would shatter under the load). Another possible early supercar might be the 1912 Bugatti Black Bess. This was a 5-litre monster with a chiselled front end and elongated rear to cheat the wind. There are other progenitors to the latterday modern supercar using different refinements of the same primitive petrol-powered approach, but they all had one goal in common behind their science: *to go faster*. These cars were, indeed, "super".

"Arguably the first supercar was launched in 1901 – the 35bhp Mercedes."

BELOW_The 1902 40bhp Mercedes-Simplex is perhaps the first "supercar" – an evolution of the science behind its 35bhp predecessor. Performance improvements were already being made at a frantic pace early on.

CHAPTER 2
STYLE VERSUS SUBSTANCE

The First World War stopped the globe in its tracks for four years and the conflict's devastating impact was felt for many more.

Introduced in 1908, Ford's Model T was the first mass-produced automobile. Here, dozens of cars are lined up ready for delivery in 1925.

This certainly applied to the world of high-end car design – while the conflict accelerated scientific innovation in many areas, luxury or high performance cars were understandably not seen as a priority, either during the war or in the gruelling period of recovery afterwards. Consequently, there were far less innovations between the wars than there were in the prodigious formative years of 1898–1914. In fact, the period 1918–1938 is not a high watermark for the science of supercars, for a number of reasons.

War and financial hardship focused people on what was *necessary*, rather than what was enjoyable; for the majority, the pursuit of speed was secondary to a much more pragmatic role for the motor vehicle. By 1918, people had become used to the sensation of driving, the novelty had worn off to some degree and this 20-year period would prove to be more about achieving "motoring for the masses" than it was about scientific excellence for the elite few.

Pushing scientific boundaries in cars has always been, and will always remain, an expensive venture and was not therefore well-suited to a war-ravaged world where many people were struggling to even put food on the table. As well as financial

RIGHT_Bentley 4.5 Supercharged, 1929

26 // Style versus Substance

ABOVE_The Bentley Speed Six was one of several hugely successful, powerful racing cars from this era. It is seen here at the finish of a six-hour race in 1929.

austerity, in the immediate aftermath of World War I there was also a shortage of raw materials. More militant workforces had led to a growth in industrial disputes, which further stifled progress. Even if a car manufacturer had sufficient funds and industrial capacity, high performance and luxury cars were frequently frowned upon, seen by many as a flagrant promotion of speed, social extravagance and financial profligacy.

This was in sharp contrast to the "man on the street" – the war had created a pent-up demand for everyday cars, by virtue of the many destroyed factories and also the lack of car production for four years. Many war-time factory owners now found themselves with huge capacity, but no more military orders. The car industry thus acquired an intense preoccupation with *volume*, as it tried to capture what was clearly a huge untapped global market. The appeal of selling a handful of fast

cars was dwarfed by the commercial potential to cash in with a mass-market car "for the people".

This muted demand for purely fast cars would only get worse as the Roaring Twenties crumbled into the Great Depression. Even the high society parties and glitzy *Great Gatsby* elite offered little demand for supercars – speed often came second to luxury and comfort with the sheer opulence of cars such as Duesenbergs and Rolls-Royces, grand tourers that made the right kind of statement at all the best dinner parties. This period was more about "show" than "go" (a proliferation of open-top cars endorses this). This was a world more concerned with high society than high science.

The end of the thirties offered little new promise: the suffocating financial crisis of the late twenties and early thirties extinguished both customer demand and engineering evolution in the upper echelons of car manufacture. Global economic frailties were largely to blame for eight-cylinder engines being replaced by a fashion for smaller, essentially cheaper six-cylinder power units, hardly a development that helped the science of supercars push forward. In Garet Garrett's 1952 biography of Henry Ford, the author says that in the aftermath of the Great Depression, "the song of the wild wheel died suddenly".

One undeniable benefit of mass-market focus was that during this period the layout of the car was pretty much standardized: a pressed-steel chassis frame; engine at the front, drive sent to the rear; leaf springs; a universal accelerator/brake/clutch pedal sequence; laminated windscreen glass; central handbrake and gearstick levers – these elements all became staple ingredients of standard road cars and were mirrored in the majority of faster vehicles, too.

The development of road cars during this time towards mass-market and affordability was in sharp contrast to matters on the race track which, from the perspective of supercar science, kept the flame alight. The twenties was a decade when many key motor sport events first took place: in 1925 the inaugural Grand Prix World Championship was held (it was won by Alfa Romeo); the first German Grand Prix came in 1925; the first British Grand Prix followed in 1926; the Nürburgring was opened in 1927; Scuderia Ferrari arrived in 1929. Brooklands would continue hosting races from its reopening in 1920 until the outbreak of war in 1939. Until 1925, speed events and hill climbs were held on public roads and beach races were still very popular. The advent of "touring car" races kept a closer affiliation with road cars as Grand Prix cars became ever more advanced.

While America led the way with a fixation on mass-market production through the twenties and thirties (and subsequently only contributed marginally to the evolution of "super" cars), it was left to the likes of Germany, France and Great Britain to fly the high-performance flag. State-subsidized racing teams from

"Many key motor sport events first took place in the twenties."

Germany and Italy helped the advancement of science in the upper end of car evolution. European manufacturers dominated Grand Prix racing and this in turn pushed the science of their fastest road cars ever forward (the opening of the German Autobahn and the Italian Autostradas only served to encourage the more speed-hungry car owner, too).

The twenties was the era of the "gentleman racer", which would prove absolutely crucial to the science of the supercar. These were often high society individuals with the finances and spare time to pour money into the development of their vehicles for racing. There was a surplus of pre-war racing cars and many monied gentlemen bought these up and modified them for their own track purposes.

Who were these racers? While pioneers such as Henry Ford were obsessed with mass-production and making the car affordable to the everyman, there was a relatively small band of eclectic, inspired and at times even eccentric individuals who had a different, much speedier agenda. Some of these gentlemen racers were wealthy officers returning from the war who didn't want to be chauffeured around in beautiful grand tourers; they wanted to be behind the wheel of a race car.

BELOW_The Bugatti Type 35 was the first car with lightweight alloy wheels, a sign that manufacturers were starting to look beyond horsepower to improve performance.

Bugatti's official historian Julius Kruta is an expert on the early and formative years of the motor car and is highly qualified to comment on the period when supercars were only really just starting to germinate: "These men wanted to have control themselves. They wanted to drive because everything else in their life was done for them. For them, driving was a sport and they felt much cooler getting out of a car themselves than being driven and the door being held open for them like royalty. They didn't want that.

"They had this romantic idea that their car was so fast that they could go and race it. They could drive on public roads to the races, park up at the circuit and change the fuel, pull a few parts off, put fresh racing tyres on, put different spark plugs in and do the race, then immediately afterwards simply drive home again for a triumphant shot of brandy and a fine cigar. That in the true sense is a 'super' car."

"The twenties was the era of the 'gentleman racer'."

Throughout the twenties, this cult of the gentlemen racers advanced the science of race cars, and by association road cars, significantly. These charismatic individuals generated a raft of writing on each of them, but for the purposes of this book, the likes of Ettore Bugatti, Ferdinand Porsche, Enzo Ferrari and the so-called Bentley Boys are the obvious headline grabbers. As previously mentioned, Enzo Ferrari started his racing team, Scuderia Ferrari, in 1929 and thus created a cult and legend that thrives to this day. Bugatti focused on racing, engineering and aesthetics, and was also one of the first to recognize that weight is the enemy of the fast car. Ferdinand Porsche had already proven his worth with the Daimler 80 PS and also his aforementioned Lohner-Porsche Mixte Hybrid way back in 1898.

Bentley was at the forefront of this era of Gentlemen Racers. Countless racing records and wins saw their cars set new benchmarks for speed and luxury. Their vehicles were heavy, yes, but that didn't stop the marque dominating Le Mans in the twenties. The infamous Bentley Boys are the most obvious example of Gentlemen Racers: cigar-smoking, spirit-drinking playboys including a pearl fisherman, a diamond magnate and various high-profile financiers. It is the Bentley Boys that also offer a neat summary of why these racers proved so influential in the evolution of supercar science, specifically because of their use of what is regarded by many automotive historians as the single most important innovation of fast cars in the inter-war years: the brutal simplicity of the supercharger.

"Bentley set new benchmarks for speed and luxury."

BELOW AND OPPOSITE_The legendary Bentley Blower, with a massive supercharger fitted to the front of the car to boost power – a genuine sledgehammer performance car.

Simple physics tells us that unless different materials are used, the bigger an engine becomes, then the *heavier* it will be. The vogue for ever-bigger engines during the twenties would have ultimately led to cars becoming impractically massive. Engineers knew that another way to boost power was needed and no other car better typifies this quest than the magnificent Blower Bentley.

Basic compressors had been around since the middle of the 19th century. An American race car had used a supercharger as early as 1908 and the first series-produced cars with this technology were two Mercedes models of 1921. Chadwick, Mercedes and Fiat had all experimented with the science of superchargers. However, the startlingly beautiful Blower Bentley is perhaps the most famous supercharged car of them all. The Blower was a 4.5-litre Bentley that appeared in 1927 with a huge supercharger slapped on the front, generating a massive power boost to catapult the car to great speeds. The increase in power came at a cost – the Blower was massive, very heavy and aerodynamically very basic. Essentially it was a sledgehammer being smashed through the air by massive force. Other contemporary cars that utilized superchargers include, but are not limited to, the Fiat 805, the Alfa Romeo 8C, the Type 55 Bugatti and the Mercedes-Benz SSK.

THE SUPER-CHARGER

A supercharger is essentially an air pump that forces more oxygen into the engine cylinders (known as "forced induction"), which then burns and creates a bigger explosion, generating more power. This is in contrast to a so-called "normally aspirated" engine, which just uses the demands of the engine at atmospheric pressure to draw in air.

On a car such as the Blower Bentley, there is a supercharger between the engine and the carburettors. This supercharger is being driven by the crankshaft. The rotation of a pair of meshing rotors creates increased suction

and this demands more air and fuel from the carburettors – essentially force-feeding the engine. This compressed fuel and air mix is then forced into the cylinder inlet ports under pressure (around one bar over atmospheric pressure) and then subsequently into the cylinders.

There are limitations to supercharging air (as with turbochargers – see page 84), specifically temperature increases due to pressure. To alleviate this, intercoolers were later introduced (not on the Bentley) to maintain supercharger and turbocharger engine efficiency.

Superchargers are not as efficient as turbochargers due to the power consumed in their operation. The mechanical power required to drive them is taken from the moving parts of the engine, namely the crankshaft. Turbochargers have (to an extent) free power supply in the form of exhaust gases.

Bentley 4.5 litre supercharged "Blower"

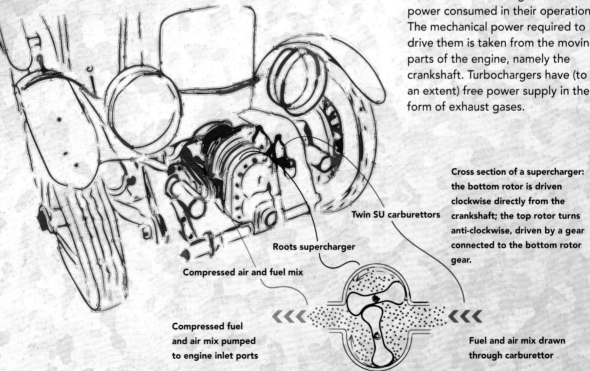

Twin SU carburettors

Roots supercharger

Compressed air and fuel mix

Cross section of a supercharger: the bottom rotor is driven clockwise directly from the crankshaft; the top rotor turns anti-clockwise, driven by a gear connected to the bottom rotor gear.

Compressed fuel and air mix pumped to engine inlet ports

Fuel and air mix drawn through carburettor

Aside from the supercharger, there were other innovations during the inter-war years, as race car manufacturers started to focus on technology, rather than just the power of the engine. Other notable innovations and refinements such as front-wheel drive, double overhead camshafts, multiple valves and gearbox synchromesh all appeared and were soon adopted by the more expensive end of motoring. The Bugatti Type 35 had the world's first alloy wheels; Bugatti's Aérolithe concept car had magnesium body panels, which would melt under welding heat so had to be rivetted; a bespoke Mercedes SSKL – the L stands for lightweight – won the German Grand Prix in 1931, complete with weight-saving holes drilled in the chassis; the Alfa Romeo 8C Tipo B Monoposto used lightweight aluminium and other alloys from the aviation industry and even had phosphor bronze valve seats; the Chrysler 300, Stutz Bearcat, Duesenberg Sis and Vauxhall 30/98 also pushed the scientific envelope of fast cars in a number of ways.

"Race car manufacturers started to focus on technology, rather than just the power of the engine."

BELOW_The beautiful sweeping lines of the 1935 Bugatti Aérolithe concept car, which significantly influenced the design of the iconic Type 57SC.

73276

As well as the obsession with power and the engine, the racing world also fuelled a fascination with one of the supercar's most dark arts: aerodynamics. This field of design was born at the end of the 19th century through a combination of experimental work by scientists such as Lilienthal and Langley, the Wright brothers and also theoretical developments by the likes of Kutta, Joukowski, Prandtl and others. By the twenties and thirties, car designers were beginning to realize that lessons could be learned from aviation and other industries to make their machines quicker and more efficient. Even so, very early aerodynamics had a slightly different motivation. Before World War II, aerodynamics was used by most manufacturers to prove that cars could be a swift means of daily transport, rather than to directly appeal to speed freaks or supercar buyers. It was a more practical search for speed than in later years.

In the early days of motor racing, regulations limited the bore of an engine, but not always the stroke, which encouraged designers to make taller and taller engines. This was an era when, quite literally, most fast cars stood tall. The majority of the great cars of the so-called vintage era (classed by expert Eric Dymock as 1920–40) all had big engines: the Vauxhall 30/98, the Lorraine-Dietrich, Alfa Romeo RL and of course the Bentleys.

Yet the brains behind the racing teams and faster marques

ABOVE_Augusto Tarabusi in his Alfa Romeo RL Sport at the 1922 Targa Florio, driving a car that typified much of the taller engine, high-performance machines of the vintage era.

TWIN OVERHEAD CAMSHAFTS/ MULTIPLE VALVES

A camshaft is a bar along which are positioned lobe-shaped cams. As the camshaft rotates, these cams open and close the valves in each cylinder. Double (or twin) overhead cams (DOHC) improve high-performance engines by allowing the positioning of inlet and exhaust valves as well as the spark plug to occupy optimized positions in the cylinder head. Combining multiple valves in conjunction with a twin overhead cam produces optimum fuel flow into the cylinders.

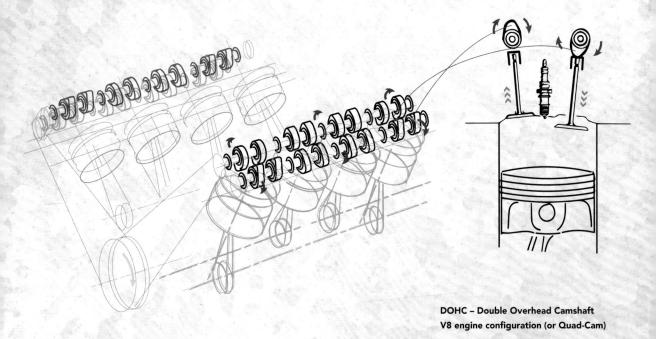

DOHC – Double Overhead Camshaft
V8 engine configuration (or Quad-Cam)

GEARBOX
SYNCHROMESH

Synchromesh, as the name suggests, enables the different gear cogs (known as "ratios") to be selected, synchronized and meshed together in harmony. Using a series of splines, bearings, selector forks, shafts, cones, hubs, sleeves and rings, synchromesh makes the gear changes smoother and faster, creating better performance and also greater driving comfort.

Selector fork in neutral position with engine idling, i.e. torque being transmitted through input shaft to countershaft only.

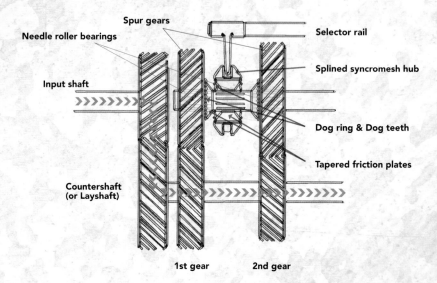

Spur gears

Needle roller bearings

Input shaft

Countershaft
(or Layshaft)

Selector rail

Splined syncromesh hub

Dog ring & Dog teeth

Tapered friction plates

1st gear   2nd gear

Selector fork moves to engage 1st gear, clutch disengaged transmits drive through output shaft to wheels

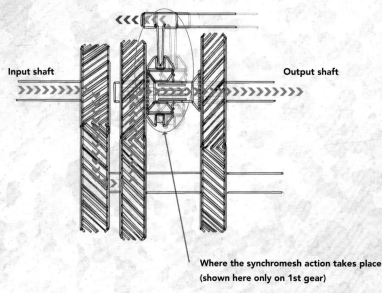

Input shaft

Output shaft

Where the synchromesh action takes place
(shown here only on 1st gear)

With clutch depressed engine torque is no longer transmitted through the input shaft. Through movement of the manual gear lever, the selector fork pushes the splined (and driven) synchromesh hub towards free-wheeling (spur) gear and takes with it the tapered friction plate and dog teeth. These mesh with the dog ring and corresponding friction plate, which then bind together (by friction), enabling the (now locked to shaft) spur gear to transmit drive via the output shaft to (in the case of rear wheel drive) a propeller shaft, differential, driveshafts and finally axles/hubs/wheels. The same process happens for all gears as the car requires engine speed that matches road speed.

>>>> Movement of synchromesh parts
>>>> Engine torque disconnected from drive
>>>> Engine torque transmitted through gearbox to wheels

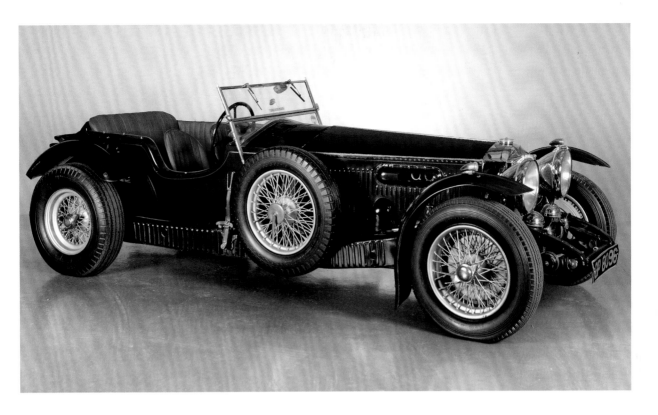

instinctively knew that this approach was potentially self-defeating, as the logic of ever-bigger engines was beginning to offer diminishing returns. Therefore the cleverest high-performance car designers started to look at making their vehicles lighter and more aerodynamic. This was the very beginning of designers manipulating elements such as the centre of gravity, chassis weight and aerodynamics.

"High-performance car designers started to look at making their vehicles lighter and more aerodynamic."

With regards to car design, many aeronautical engineers moved over into the automotive industry and their detailed knowledge proved instantly beneficial in the science of high-performance cars. For example, Paul Jaray had previously designed Zeppelins for military use, but soon became renowned for creating many streamlined car designs. He included elements such as smooth body panels, integrated headlamps and curved windshields, all designed with aerodynamics in mind. Jaray was also one of the first to use mathematics in car design, creating formulas that attempted to decode how much power an engine would need to overcome air resistance. He even set up a base in the USA and enjoyed considerable transatlantic success with his company, the Jaray Streamline Corporation.

There are other very notable examples of early aero. The Invicta 4.5-litre S-Type Low-Chassis Tourer was a touring car, rather than a "super"car, but it caused a sensation on its launch in 1931, despite the impact of the Great Depression hanging heavy over the world's monied elite. The low-slung chassis was a revelation and is considered by many automotive historians to be a watershed moment in car design. Its innovation is highly relevant to the science of the supercar as the designers specifically positioned the axles above the frame rails instead of below, in order to achieve a much lower centre of gravity.

The Bugatti Atlantic 57SC could certainly claim to be one of the very first supercars that could slip through the air more efficiently than its rivals. The 57SC's predecessor, the Type 57, was a success on the race track, even winning at Le Mans having lapped the rest of the field. In 1936, following that historic triumph, Jean Bugatti produced the stunning 57SC, which seamlessly matched aerodynamic flair with design genius. Bugatti added a supercharger that created over 200bhp, and made the car as low and light as possible. This perfect blend of art and science remains an icon of design and art to this day – one example of the 57SC Atlantic became the most expensive car ever when it sold at auction for over $40 million in 2010.

Bugatti historian and vintage car expert Julius Kruta generously cites other examples of very early cars that recognized the need for more than just sheer force to go faster: "Aside from the Bugatti 575, you had the Invicta 4.5-litre S-Type, the Bentley Speed Six and the Mercedes S-Type and a couple of other cars where ground clearance was lowered to refine the centre of gravity. They were very, very fast cars. They were

> **"In 1936, Jean Bugatti produced the stunning 57SC, which seamlessly matched aerodynamic flair with design genius."**

all doing about 200kmph [about 124mph], although they had very little room inside. You could use them for long distance travelling but you rarely did, I would say. They were the pinnacle of design."

Other similarly lithe and sinuous cars from the period reveal the emerging science of aerodynamics beginning to filter into car design, vehicles such as the Teardrop Coupe Talbot-Lago T150C-SS, the Delahaye 165 and Alfa Romeo's stunning 28C 2900. The Pierce Silver Arrow was another stunning looking car. Other examples of extreme focus on aero might include the tank-like Bugatti Type 32, the BMW 328 of 1937–9 and the futuristic 1932 Auto Union Type C. It is important to note that – as with most supercars – these were low volume cars: only 30 Alfa Romeo Type 28C 2900s and a mere four Bugatti 57SCs were made.

However, these beautiful cars were still the exception to the rule. According to vintage car expert Eric Dymock in his book *Postwar Sports Cars: The Modern Classics*, even towards the end of the thirties fast cars were still relying predominantly on power for performance. Further, he suggests that those cars that were lowered usually still boasted a contradictory excess of "goodies" on the bodywork, such as wide wing mirrors and other trinkets. Similarly, road holding was "curiously neglected" while suspensions were outdated and seen as a tool for comfort, not handling; gears remained large and cumbersome; hydraulic brakes were in their infancy and much of the science of these faster cars was essentially the same as at the start of the thirties. There was still much work to be done.

CHAPTER 3
EMERGING TECHNOLOGIES

If the thirties is not regarded as a golden period for the fast car, then the final insult of a tough ten years was that just as the post-Depression world was hauling itself to its feet again, World War II started.

Production of anything luxurious, extravagant or not "essential" shuddered to a halt. Supercars, or at least high-performance cars of the day, were effectively made obsolete overnight. However, the notion that "necessity is the mother of invention" is crucial to this bleak period of world history and supercar science. Car factories all around the world were repurposed for the war effort (to make tanks and weapons, even tin hats, pots and pans). Vehicles originally meant for the public were requisitioned for military use and fuel rapidly became so scarce and heavily rationed that extensive private transport was at times impossible. Ambitious plans for sports cars, convertibles and luxury vehicles were hastily scrapped. Necessarily, the skill set of the engineering brains behind these performance cars was also redirected, at least for now. The development of the road-going car all but stopped for the six war years.

Even when war was finally over in 1945, the recovery would prove to be stuttering and slow. Re-starting the automotive industry was not an overnight task because so many factories had been destroyed or severely damaged. A shortage of steel

RIGHT_Jaguar XK120, 1949

also led to manufacturers looking for lighter materials (the 1946 J1 Allard being one example). Immediately post-war, most new cars were simply re-hashes of pre-war designs, as little time or money was available to develop a new car from scratch – this stifled everyday cars and certainly hindered any initial plans for new performance cars. While America focused on big, gas-guzzling coupes and there was a glut of smaller British sports cars such as MGs, Triumphs and Healeys, Europe tried to tighten its ration-era belt with punitive taxes on big and high-performance vehicles, instead promoting the smaller, more utilitarian car. In his definitive *The Classic Car Book*, Giles Chapman cites the flurry of "people's cars" that this environment encouraged, including the Morris Minor, Renault 4CV, Citroën 2CV, Fiat 500C, Saab 92 and, of course, the Volkswagen Beetle. The design culture was for frugal, not sensational, economic not exhilarating and accordingly the science evolved for that purpose. Delmar Roos, the chief engineer of US car-maker Willys-Overland, had previously said that "the object of the automobile is to transport a given number of people with reasonable comfort, with the least consumption of gasoline, oil and rubber, and for the slightest operating cost and prime price". Not exactly a manifesto for the future of the supercar …

> **"Most new cars were simply re-hashes of pre-war designs."**

So where did this leave high-performance vehicle design and the science of the supercar? The post-war years marked a slow-burning recovery period that saw only modest innovation in most areas of high-performance car science and design. Improvements across the automotive industry with suspension, brakes and tyres were more incremental than seismic, and with the modern supercar not yet in existence, the era is a rather nebulous and vague moment in the history of the genre.

However, all was not lost. From a scientific point of view, the black hole of tragedy that had been the war had in fact compressed decades of development into a few intense years and created a generation of super-engineers – brains that were accustomed to their genius being put under the most intense scrutiny and time-sensitive, potentially life-or-death pressure. In Germany, engineers were explicitly forbidden from building aircraft after the war, so many of those great minds turned to car design instead. On the Allied side, similarly inventive minds exited the war effort and were snapped up by a car industry eager to soak in their expertise. (A similar migration across engineering fields saw a post-war boom in motor racing, utilizing now-redundant airfields hungry for income and purpose; this effectively spawned the modern UK motorsport industry, complete with all its ancillary car development industries, which ultimately became a world-renowned hub of motor racing.)

ABOVE_A special exhibition of the iconic "People's Car", the VW Beetle, in West Berlin in July 1949.

RIGHT_An ultra-rare right-hand drive Aston Martin DB4 GT, a car much loved in America and the epitome of English style.

There were moments of inspiration – the 1948 Jaguar XK120 then the C-Type and latterly the D-Type that dominated Le Mans were all hugely invigorating (the XK120 boasted the first twin-cam engine in series production). As well as Jaguar, British cars such as the Aston Martin DB4 sold well overseas as a booming America bounced back better than Europe – helped by the fact that, unlike most of their European automotive counterparts, many US factories remained intact. Two historic continental marques launched onto the world of the performance road car: Ferrari's 1949 Ferrari 166 Inter was the first road-going purpose-built sports car from that famous marque, and perhaps predictably enjoyed some racing success with numerous variants, too (many argue that its V12 was the foremost example of that form at the time); meanwhile, Porsche's dainty 356 signalled that brand's entry into the world of the sports car.

From a scientific point of view, the fifties saw the introduction of a unitary body rather than coachwork that was bolted on top of a chassis. Previously, for example, Ferrari initially only sold 166s in chassis form, which were then bodied to the customers' desires. This was not necessarily a negative – the stunning

Cisitalia 202 remains a watershed moment in automotive design with its sinuous and flowing bodywork. However, the vogue post-war was to produce a car with a "standard" shell already in place.

The fifties and part of the sixties also saw a great increase in power, as manufacturers continued to search for ever better performance. War-time developments in aviation engines certainly helped them in this quest – in ten years, the top speed of fighter aircraft had doubled from 300mph to 600mph. Other helpful progress, such as the advent of high-octane fuel, had also boosted performance across all mechanical ventures. However, the science of these fast cars would evolve much more as a result of the work of innovative designers who strived for more than just greater power.

> "As well as Jaguar, British cars such as the Aston Martin DB4 sold well overseas."

As the world gradually recovered, a new class of wealthy elite began to look for ways to spend their money. As ever, the rich looked to the car as one of the ultimate status symbols and it was in 1954 that a vehicle arrived that many regard as *the* first ever

LEFT AND OPPOSITE_
With a stunning
spaceframe chassis
and numerous
other innovative
technologies, the
Gullwing makes
a strong case for
being the very first
modern supercar.

ABOVE_An original
German advert
for the Mercedes
300SL Gullwing,
launched in 1954.

modern supercar: the Mercedes 300SL Gullwing … and it had
the science to back up the hype.

The Gullwing's story reflects the society of the time.
Renowned American Mercedes dealer/distributor Max Hoffman
had convinced the company that a road-going version of their
hugely successful SL race cars would sell well in the US. The car's
scientific and developmental genesis came from the engineering
brain of Mercedes' chief race engineer, Rudolf Uhlenhaut. With
the company dominating on the race track, there was a clear
opportunity to capitalize on this success with its road cars.
Insufficient post-war funding and the bombing of most Daimler-
Benz factories meant that Uhlenhaut had to work from the basis
of the well-regarded 300 Sedan, rather than start a new car from
scratch. He would cannibalize a majority of the 300's engine
as well as the four-speed manual transmission and suspension.
However, beyond that, the new car's science was revolutionary.

The Gullwing had what was known as a "spaceframe chassis",
namely a lightweight, strong chassis constructed from a complex
maze of ladder-like frames of tubular steel. Much of the science
came from the aviation industry, where tubular frames had
been commonplace in many military and aviation applications
for some time and had evolved hugely during the war years.
So it was not a great leap of faith to transfer this innovation to
high-performance cars. Further, the art of structural welding
had progressed through the war to a point where potential
breakages and structural fragility was a much less worrying
issue. Other marques such as Ferrari, AC and BMW had realized
that a stiff chassis and supple springs would assist handling and
cornering, but it was the Gullwing that took the biggest step
forward in terms of science. The triangulated framework was
a masterpiece to weld and presented a significant obstacle for

The pioneering science did not stop there. The Gullwing also introduced an independent four-wheel suspension, improving the handling compared to its rivals (still rudimentary by modern standards but a scientific evolution at the time). The car was also the first road-going vehicle to successfully use direct fuel injection, another idea culled from aeronautical research.

A handful of early Gullwing models were made using all-aluminium skins but this was later changed to mainly steel (on both variants, the SL stands for Sport Light). The science of the car also dictated that the engine be canted at 50 degrees to the left to keep the centre of gravity lower and the overall height of the car under control with a lower driving position – another sign that supercar engineers were looking at ever more comprehensive solutions in the search for greater performance. The Gullwing was the quickest car in the world at the time of its launch in 1954 at the New York International Auto Show and proved Hoffman's commercial instincts right by going on to become a huge success, despite its price being ten times that of a normal family car and double the cost of a Jaguar D-Type. Understandably, the Gullwing's pioneering science, stunning looks and world-leading performance singled it out for many as the first ever modern supercar.

BELOW_The Gullwing was so precise in its detail and so well-engineered that it made other sports cars look dated overnight, and still produced race-winning performance.

Just as the supercar genre had seemingly been ignited by the Gullwing, the 1956 fuel shortage provoked by the Suez Crisis delivered a sharp shock to the global system and luxuries such as supercars faltered once more. The vogue for smaller, economical cars and affordability still dominated. Then, a British car company launched a car that caused an absolute sensation and indelibly marked the science of supercars forever. That car was the Jaguar E-Type, also sometimes known as the XK-E. Even though some observers regard the E-Type as a sports car, not a supercar, its impact on the science of supercars cannot be denied. Launched early in 1961 at the Geneva Motor Show to a stunned world, the stylish and relatively affordable Jaguar XK120 had set the scene for the E-Type to follow and its successor made many contemporary cars look instantly old-fashioned. The science stands up for itself: 400lbs lighter than its predecessor (the XK150); half the price of a Ferrari 250GT but with the same top speed and ten percent more power; perhaps most crucially, the aerodynamics were a central element of the design and subsequent appeal.

Michael Quinn is the grandson of Sir William Lyons, the co-founder of Jaguar, making him uniquely placed to comment on the E-Type; he has some fascinating opinions on the legacy of this remarkable car: "The obvious discussion is always going to be about the looks, the exterior, but actually the car had

OPPOSITE_Regarded by many – including Enzo Ferrari himself – as the most beautiful car ever built, the Jaguar E-Type redefined the marriage between high performance and scintillating good looks.

innovative independent rear suspension and the subframe was a development of the D-Type race car, designed to continue the success at Le Mans, which was influenced by aeronautical science, both of which were ahead of their time. Compared to the purity of the D-Type, of course, the E-Type design was slightly compromised because they had to have an eye on the practicality and commerciality of the end vehicle, with such things as interior space, lights, bumpers and so on.

"By modern standards it isn't particularly aerodynamic but for its time it was pretty impressive. I suspect the aero on the E-Type wasn't as prioritized as it was on the D-Type, simply because one was a racing car and one was for commercial sale. The aesthetics inevitably took over a little and I am sure my grandfather would have frustrated designer Malcolm Sayer a little when pushing for changes to make the car look prettier.

"The impact of the E-type on the science of supercars cannot be denied."

"It is true that they tested the car's aero using tufts of wool. Sayer would place them at certain parts of the car and then the test driver Norman Dewis would drive at speed while Malcolm would observe from another vehicle what the tufts were doing, which was a very simple but effective visual sign of where the air was moving. Similarly Dr Samir Klat employed the same principle when developing the low drag E-Type Lightweight race car."

Michael also explains how the team was completely taken aback at the impact and huge success of the car, which in

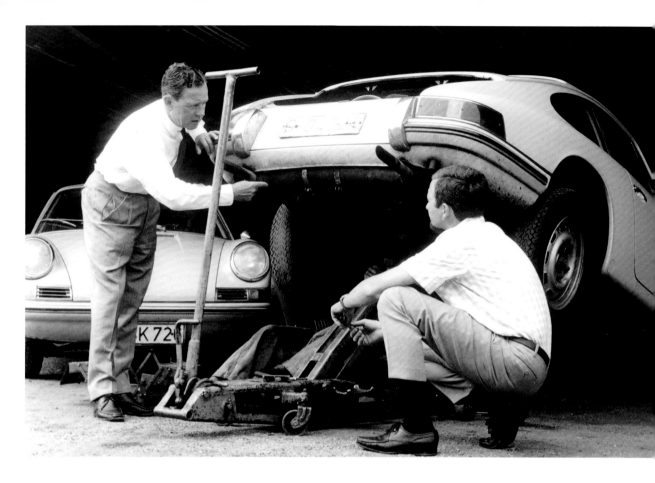

ABOVE_Ferdinand
Porsche and his
son working on a
901/911 prototype,
the genesis
of a car that
has stubbornly
remained as the
exception to the
mid-engined
supercar rule.

itself created a scientific challenge behind the scenes: "For
the second time they completely underestimated the impact
of a new model. The same had happened back in 1948 with
the XK120. With the E-Type, they were suddenly faced with
creating the tooling and manufacturing capability to produce this
incredible car at a very low price. That created a whole new set
of challenges. My grandfather was incredibly tough on suppliers,
he ran a very tight ship, very efficient and constantly looking
to keep costs down. He had a team of fantastic, superlative
engineers who created absolutely the right product at the right
time for a remarkable price." History shows that the resulting
E-Type pushed the boundaries of high-performance car science
as well as the engineering infrastructure to back that up.

Aside from the two pivotal scientific moments represented by
the Gullwing and E-Type, most high-end car manufacturers were
relatively slow to innovate – often concentrating on luxurious
saloons rather than purebred fast cars. Nonetheless, the science
of supercars was beginning to accelerate. The period after
the war saw the very earliest fibreglass bodies being used, as
lightness became a new holy grail. The 1953 Corvette made one
of the very first examples, and in the UK the influential English
designer, racing team owner and motorsport pioneer Colin
Chapman's 1957 Lotus Elite Type 14 flew the flag for lightweight-

bodied cars, even if neither would be regarded by most as pure supercars (this was the first car constructed from glass reinforced plastic; notably, this was also the first car to carry a full monocoque, an achievement mirrored in F1 by the Lotus team's Type 25).

One hugely significant post-war development that did fuel the science of innovation for decades was the close relationship between racing cars and supercars. Motorsport quickly reawakened after the end of the war. Although Brooklands was initially still being used by the aviation industry, many redundant airfields around Europe became playgrounds for professional and amateur racers who suddenly found an abundance of locations to express their need for speed. Manufacturers had been making "road legal" versions of race cars as well as actual race cars that could be used on the road for some time. The Ferrari 375MM, the Ferrari 250GT and the Porsche 550/500RS are all such vehicles. The mid-fifties series of Ferrari "superfasts" fuelled even more interest in the idea of road-going cars that could perform almost as well as a race car. For their early cars, Ferrari were unabashed about the influence of race car engineering in their road car science. Similarly, Jaguar created the 1957 XKSS out of the remainders of the legendary D-Type. As the post-Gullwing supercar world evolved into the sixties and beyond, the pioneering world of motor racing would play a central part in the evolution of the science of the high-performance car for the road, as the commercial relationship between racing and road cars was at its most beneficial.

For example, the post-war period saw a leap in developmental testing of the science behind these new fast cars. Rigs were designed for the first time to test cars for endurance, waterproof ability, soundproofing and so on. A perfect example would be the use of the wind tunnel, a sealed, controlled room (test section) into which is placed a scale model while air is blown across the space at a controlled speed to measure the three basic forces of lift, drag and side force. The science was not new – the first wind tunnel actually pre-dates the first manned flight by 30 years. Once again, it was science culled from aerospace – according to NASA, Frank H. Wenham, a Council Member of the Aeronautical Society of Great Britain, is credited with designing and operating the first wind tunnel in 1871. By 1948, the Healey 2.4-litre was the world's fastest production road car on its launch – and was developed with the aid of a wind tunnel.

Overall, the first two decades after the war mostly saw only incremental improvements in car development. With the Gullwing and E-Type among a few notable others, there were nonetheless encouraging signs that the science behind the supercar genre was finally beginning to explode into its modern form. However, in 1966 a small manufacturer in a village in northern Italy launched a car that, from a scientific point of view, changed *everything*.

"Motorsport quickly reawakened after the end of the war."

CHAPTER 4

THE MID-ENGINED MACHINE

The strong global economy of the first years of the sixties meant there was money to be made with a new supercar.

OPPOSITE_The Lamborghini Miura, the first of the legendary mid-engined supercars and a machine that redefined the science of the entire genre.

For many car enthusiasts, that decade was a golden era when roads were not yet constantly congested, in some cases speed limits did not apply and cars with great handling offered unbridled joy for the skilled driver. Family cars enjoyed record sales but so too did sports cars, as people looked for exciting ways to spend their spare cash. The first generation of post-war car buyers had also come of age so the market was ripe for exploitation. What did this interest and demand mean for the science behind the high-performance car?

Essentially the Year Zero for the science of modern supercars can be traced back to the 1966 Lamborghini Miura, specifically its so-called "mid-engined layout". Until the mid-sixties, engines in most cars predominantly remained in the front. Likewise, the layout of fast cars replicated that of regular automobiles in many respects – with long engines crammed under elongated bonnets. However, this places a huge amount of weight at one end of the car. Racing cars were the first to experiment with the idea that road holding and performance might be improved if the engine was further back in the vehicle. These race engineers were not actually the first to acknowledge the importance of layout – as far back as 1900 the NW Rennzweier racing car had a mid-engine, rear-wheel drive layout. There was a Bugatti Type 251

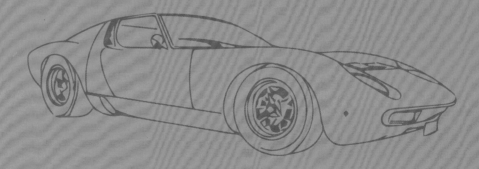

RIGHT_Lamborghini Miura, 1966

ABOVE_A 1954 Porsche 550 Spyder on the East River waterfront in Queens, New York. James Dean famously dubbed his Spyder "The Little Bastard", and would eventually lose his life in the vehicle.

racing in Grand Prix in 1956 for a short time that was transversely mounted. Ferdinand Porsche pushed the envelope further by experimenting with the mid-engine layout in designs for the Auto Union Grand Prix cars of the thirties and the Porsche 550 Spyder of the mid-fifties. Then he turned logic on its head (or end) with the 1963 Porsche 911 model, which defiantly planted the heavy engine right at the rear of the car. It would take Porsche many decades of refinement to master what initially proved to be a tail-happy layout, and one that persists stubbornly to this day as the exception that proves the rule.

It was the sixties that saw the mid-engined approach take hold in both racing and road cars. On the track, pioneers such as John Cooper (of the Cooper Car Company), Jack Brabham, Colin Chapman and Lola Cars unleashed the revolution. Numerous F1 cars and endurance racers used the idea, such as the Ferrari 250 LM. In 1963 the Italian company ATS (Automobili Turismo e Sport) had unveiled the spaceframe ATS 2500 GT, which was a mid-engined high-performance car, sadly often overlooked in the history books as the project collapsed under financial strain.

If it was the English who headed up mid-engined science in racing, it was in Continental Europe that the biggest jump was made for road cars. In the world of supercars, it was a man who had made his fortune with tractors, air-conditioning units and gas heaters that made the leap of faith before all others: Ferruccio Lamborghini. The son of a viticulturalist and himself a former Air

> **"In 1963, Ferdinand Porsche turned logic on its head (or end) with the 911, planting the engine at the rear of the car."**

ABOVE AND
OPPOSITE_Original
Lamborghini Miura
publicity shots
– the car caused
a worldwide
sensation on its
1965 launch.

Force mechanic, Ferruccio had bought a number of beautiful cars
with the income from his entrepreneurial endeavours, including
various Ferraris – he even competed in
the Mille Miglia in 1948. Lamborghini legend states that when
 clutch on one of his Ferraris broke, he noticed it was essentially
the same construction as that used on his tractors. Ferruccio
asked Enzo Ferrari for a replacement and – so the story goes –
Ferrari told him to stick to building tractors.

Ferruccio didn't stick to building tractors. Instead, aged 45,
he founded Lamborghini Automobili in 1961, basing himself in
Sant'Agata Bolognese, less than 20 miles from Ferrari's factory.
His first car was the touring 350 GT in 1964, but it was his
revolutionary creation of two years later that set the world of
supercar science alight: the Miura.

There has been some controversy over exactly who worked
on which stages of the Miura, but suffice to say that the design
team made the monumental decision to use a mid-engined
configuration. This might seem logical now, but back then Enzo
Ferrari was still insisting that "the horse belongs in front of the
chariot" (as indeed were pretty much all of his performance car
contemporaries). Ferruccio himself was also initially reluctant, but
when he saw the designs and listened to the science he was won
over. With no test track of their own, Lamborghini trialled the
pioneering new car on the roads around the factory.

The Miura was created as the P400 (for *posteriore* and
4.0 litre) and was the world's first mid-engined supercar. The
big 350bhp V12 naturally aspirated engine was able to fit into

the smaller road car layout because it was transversely mounted, rotated through 90 degrees, rather than running in line with the car. Also, the engine block, crankcase and transmission were all in one complex alloy casting – similar in fact to the transversely mounted set-up in a Mini. The seats were notably further forward than usual, which also benefited the front and rear overhangs, the engine bay location and positioning of the cockpit. The brilliant young designer Marcello Gandini masterminded the car and its curvaceous and beautiful shell – standing just 105.5cm (41½in) high – creating an object of beauty that is still to this day one of the finest blends of Italian flair and pioneering science.

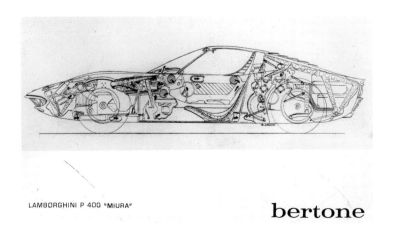

LAMBORGHINI P 400 "MIURA"

bertone

LEFT_ A cut-away of the revolutionary mid-engined Miura, the first modern supercar.

BELOW_The howl
of a naturally
aspirated
Lamborghini V12
Miura engine
remains one of the
most exhilarating
sounds of the
supercar world.

THE MID-ENGINED LAYOUT

The Lamborghini's mid-engined layout allowed all major systems to be packaged within the "wheelbase" – the distance between the front and rear-wheel centres.

By making the power plant in the Miura mid-engined, the chassis layout constrained some of the forces of physics better than rear- or front-engined cars. Essentially this created far superior handling, but two scientific principles are at work behind this simplistic statement: moments of inertia and the centre of gravity of the car.

(1) All moving cars are subject to moments of inertia. Imagine a beam sticking out of the back of a car, ten metres beyond the rear wheels, with 100kg (220lbs) attached at the end. The forces the car experiences under braking (nose goes down) and/or accelerating (nose goes up) are known as "pitch". The radial forces the car experiences when cornering are known as "yaw" and "roll".

(2) All supercar design engineers strive to reduce the centre of gravity (C of G) height. Imagine a beam with a 100kg (220lbs) weight at the end, projecting vertically upwards from the centre of the car. Now drive that car fast into a corner. Under braking, the moments of inertia through pitch, roll and yaw would cause a large variation in the car's attitude and, with steering applied, the car will experience huge instability, sufficient to ensure it will not get round the corner without incident.

The mid-engined positioning optimizes the Miura's layout and extracts the very best performance out of the car in relation to the aforementioned factors. One disadvantage was that the Miura had a propensity to understeer as the fuel tank emptied. The changing weight of the tank, situated in the front of the car, had an unwanted influence, whereby weight distribution moved rearwards as the tank became lighter. Nonetheless, compared to front- or rear-engined cars, the Miura's mid-engined layout was a revelation in terms of performance and superior road holding and grip.

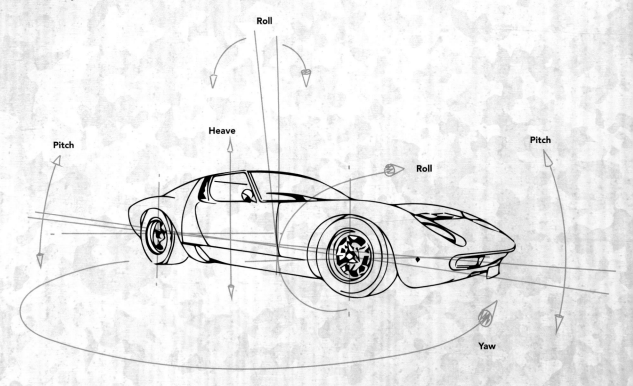

Roll

Heave

Pitch

Pitch

Roll

Yaw

The Miura also has important nods to aerodynamics and other innovations that are often overlooked in the clamour to herald the first ever mid-engined supercar. Holes were cut into the chassis to make it lighter without compromising structural integrity. A racing-style double wishbone suspension was added as well as large disc brakes. The body had air vents in the door panels and even the headlights were designed with aerodynamics and airflow as a concern.

It is a mark of how significant a moment this car was in supercar science that when a bare rolling chassis was exhibited at the Turin Motor Show in 1965 – that is to say, with no body shell – the press and public went wild. The Miura cost four times as much as an E-Type Jaguar, but orders for what was the fastest production car in the world flooded in.

History tells us that the Miura wasn't perfect: some drivers felt the mid-engined layout created handling deficiencies; there was a tendency for the nose to lift at speed; the car was always noisy with that sublime V12 engine so near to the driver's ears; there were reservations in some quarters about the sheet steel chassis and the suspension could be less than comfortable. However, those nuances were all forgiven. The fact is that supercar science changed forever the day that the Miura arrived and, for many, the famous Lamborghini represents the beginning of the modern supercar.

ABOVE AND OPPOSITE_ The devil is in the detail: the Miura's multiple style flourishes left it head and shoulders above its supercar contemporaries.

Having taken into consideration the science behind the Miura's layout, how does that manifest itself on the road? In other words, in the hands of a skilled driver, what is the impact of a mid-engined layout? Given the seismic impact of the Miura, who better to discuss this issue than Valentino Balboni, Lamborghini's legendary and former chief test driver – he held the position for over 40 years – a man so revered in supercar circles that in 2009 the famous Italian marque even named a special edition Lamborghini Gallardo after him.

"The mid-engined layout is so important in sports cars," explains Valentino, "because it is a technical necessity relating to weight distribution. With that layout the car can optimize weight, balance and the expected performance. I love the attitude of the Miura, there is a little bit of oversteer which is good, in a very rewarding way. Also having a transverse engine was even more special and different because with the block and

"The Miura's driving characteristics were excellent."

Valentino Balboni, Lamborghini's chief test driver

everything included saved a lot of space, therefore they were able to create a limited dimension car which was better performing. Every car has different attitudes and characteristics, so it depends what you want in terms of performance, but for sure a rear-engined car has more oversteer. It also depends on what each manufacturer wants to get across to the customer but also how they are looking to develop their cars compared to their competitors.

"From a technical point of view, the Miura's driving characteristics were excellent. Approaching a bend, you'd have to use a certain technique, but definitely compared to contemporary cars at the time it performed more impressively, and in the hands of a professional it is superb. The P400 and P400S can be more nervous in terms of driving, they require more driving, but the later SV with bigger tyres was more neutral, a delight. If we talk about the general performance of the Miura, I would separate the first P400 with the best of the later SV. During the development of any car, you start the project

with ideas and energy but then slowly, slowly, you realize that the first ideas can be improved. Without a doubt, the Miura wrote the first page in terms of the layout of supercars."

Seemingly overnight the Miura rendered all other supercars old-fashioned. In the aftermath of that seminal supercar, other manufacturers scrambled to get up to speed. Later in 1966, the Lotus Europa was launched with a mid-engined layout, superseding the light sports car, the Lotus Elan, which had debuted with a fibreglass body and mid-engined layout in 1962. The front-engined, rear-drive Ferrari 365 GTB/4 Daytona of 1968 seemed somewhat dated by comparison to the Miura. De Tomaso produced the cool Mangusta (named after an animal that eats cobras), then in 1968 Ferrari launched the effortlessly pretty Dino, a mid-engined, V6 junior prancing horse named after Enzo's son and heir (the car was badged as a Dino to separate it from the more elite V12 cars that Ferrari produced).

OPPOSITE AND ABOVE_In the wake of the remarkable Miura, rival supercar companies struggled to play catch up. Clockwise from top left: the Ferrari Dino, the Ferrari 365 GTB/4 Daytona and the De Tomaso Mangusta roadster.

"In the aftermath of the Miura, other manufacturers scrambled to get up to speed."

However, aside from these examples, it is remarkable to note that despite the Miura's massive scientific impact, the sixties ended with very few mid-engined supercars actually being launched. With petrol still reasonably priced and money plentiful, cars were often heavily engined and gas-guzzling, achieving performance with little subtlety – the oft-maligned muscle car phenomenon in the USA was perhaps at its peak in the latter half of the decade, for example. *CAR* magazine's *The Story of the Supercar* noted that "Swinging though the sixties may have been, in supercar evolution terms, they were near-stagnant."

CHAPTER 5

THE SCIENCE OF THEATRE

As the seventies dawned, the future did not look good for the supercar or for the funding of the science behind these exotic machines.

OPPOSITE_The Lamborghini Countach redefined what a supercar represented – best known for its stunning "space-age" looks, the Lambo was in fact also the fastest car in the world at the time of its launch.

With only the Miura and Dino flying the mid-engined flag with any success before the start of the decade, optimism in the supercar sector was in short supply. To compound this, a thirst for small family cars and the newly introduced 4x4 vehicles was attracting the most headlines and development budgets. The oil crisis of 1973 produced rocketing prices and chronic fuel shortages, while a biting global recession followed by another energy crisis at the tail-end of the decade offered little respite. Many marques folded and the British motor car industry in particular was savaged during these years. Across the Atlantic, America introduced a 55mph speed limit on highways as well as stringent emissions tests, pretty much extinguishing the muscle car genre overnight. Also, that long-standing bastion of supercar influence, motorsport, now had race cars that were hitting in excess of 200mph, so the symbiotic relationship between road and race vehicles was becoming ever more distant. This in turn meant that supercar manufacturers were increasingly having to justify the costs of a new car in its own right, without the aid of racing tech to subsidize the research and development costs. Instead of science being tasked with the question, "How fast can that supercar go?" the designers and engineers were increasingly faced with the rather less exciting challenge, "How clean, safe and fuel efficient is it?" So from the perspective of supercar science, the seventies was not a golden era.

RIGHT_Porsche 911 Turbo, 1975

Actually, on a superficial level at least, the seventies
was a decade of fantastic, extravagant supercar creations,
including perhaps the most exotic of them all, the Lamborghini
Countach. This was the era of the so-called "wedge" supercar as
manufacturers reveled in the benefit of a mid-engined layout that
allowed their cars to have a lower nose, thereby creating a host
of futuristic, space-age type designs. Apart from the Countach,
there are plenty of examples of the classic supercar wedge
from this period: it took Ferrari until 1973 to truly respond to
the Miura with the beautiful mid-engined Berlinetta Boxer 365
GT4/BB (even then critics were bemused by the engine sitting
on top of the gearbox, thus raising the car's centre of gravity);
Lamborghini had already produced the Countach's lesser sibling,
the mid-engined Urraco; the Maserati Bora was the first mid-
engined road car from that marque, followed by the Merak; the
1976 Lotus Esprit was a classic wedge shape and found fame
on the silver screen as James Bond's supercar of choice; the
brutal and striking De Tomaso Pantera remains one of the most
underrated of all supercars; even BMW chipped in with the car it
created for Group 5 competition, the frequently overlooked M1.

However, it was Lamborghini that yet again stole all the
plaudits when the design genius Marcello Gandini produced the
iconic Countach in 1974. The story of this landmark supercar has
been well documented, so suffice to say when it was launched it
made all other high-performance cars look *ancient* overnight. The
styling was drastically futuristic and the performance ballistic,
with its top speed of 183mph making it the fastest production
car in the world. Arguably the Countach remains the ultimate
poster supercar, found on millions of bedroom walls around the
world to this day. The car's name (pronounced *coon-tash* not

count-ash) is a non-obscene expletive in the Piedmontese dialect of Italian, uttered when a man sees a very beautiful woman. The story goes that this was the first reaction of a reporter on seeing a concept version of the car.

The Countach's legend is secure and it is arguable that it will never be surpassed in terms of sheer extravagance and impact in the supercar world. However, what was the *science* behind the Countach? This was a car that pushed many engineering boundaries for the time. The famous scissor doors created a supercar design icon but, in actuality, were incredibly impractical. The large sills made getting into the driver's seat a challenge. The rear visibility was almost non-existent, so it was not unheard of for Countach owners to parallel park by opening the scissor doors and sitting on the wide sill while looking tentatively over the rear of the car. Once inside, the near-horizontal driving position was certainly a challenge, even for the most ardent fan of the car. "As far as the drive is concerned," explains Lamborghini's legendary salesman Steve Higgins, "like all of these older cars, the Countach was challenging! The interior is somewhat cramped, the rear vision is minimal, it's not a car that you jump in to go cruising up and down the road, you have got to really mean it and want to do it when you get in a Countach. It is noisy and hot with a heavy clutch and stiff gear shift. But really, look at it – do you care?"

> **"The seventies was a decade of fantastic supercar creations."**

BELOW_The Maserati Bora is notable for the distinctive design of its rear window.

The famous bodywork was made from aircraft-grade aluminium, an approach culled from many race cars of the day. Underneath, an innovative tubular spaceframe showed that Lamborghini were also aware of the need for increased lightness, wherever possible.

To the layman, the sleek, wedge-shaped design suggests that the Countach must be the most aerodynamic car ever made. However, the famous Lambo is not especially drag efficient, with a coefficient figure of 0.42. The Countach was not alone in this – ironically, the classic seventies wedge look is actually almost anathema to the end goal of slippery aerodynamics and coherent, multi-purpose high performance packaging.

Adrian Newey is one of the all-time most successful designers of Formula 1 cars, with well over 100 race wins and is universally regarded as a guru in the field of aerodynamics: "When the Countach first came out, it was of course a very dramatic car without doubt, I think it did lead the way. However, its practicality and certainly its aerodynamic values are highly questionable. It wasn't the most practical of cars, but it was *the* dramatic car of the moment, if you like. But I think probably with the exception of the Countach, you could say that the seventies is a pretty dry period in automotive styling."

"The sleek, wedge-shaped Countach is not especially drag efficient."

Gordon Murray, the designer of the legendary McLaren F1 supercar, concurs, suggesting that part of the weakness of some of the seventies wedge cars is that they were designed with trends in mind as much as, if not more than, the engineering or driveability. "Trends are very dodgy in car design – many of the 'origami-type' cars are so tied to the seventies. Apart from the really outrageous ones like the Countach, most of those have lost their place now, in my view."

The seventies was therefore a period when the look of a supercar was often somewhat more beguiling than the actual science. However, delve beyond the extravagant styling and it is possible to find more innovative science emerging. It is what is under the hood of one car in particular that yields the most significant scientific step forward for the supercar: the Porsche 911 Turbo.

The science of turbos dates way back to the advent of the motor car. Daimler and Diesel investigated ways of pre-compressing the air that went into their early internal combustion engines. World War I saw some French fighter planes fitted with basic turbos, with varying degrees of success. Many cite the 1925 turbo of Swiss engineer Alfred Büchi as the first useable exhaust gas turbo, boosting his engine's power by 40 percent. Early use in large aeronautical, industrial and marine engines was followed by applications in large trucks and lorries. Rapid development in World War II reduced the size of turbos and their efficiency, so it was perhaps inevitable that their use in road cars would soon become more commonplace. Chevrolet's Corvair Monza and Oldsmobile's Jetfire option on the 1962 Cutlass had been the very first road cars to experiment with turbos in the early sixties, but reliability issues saw that primitive technology quickly dropped.

"The science of turbos dates back to the advent of the motor car."

The BMW 2002 Turbo saloon was launched in 1973, evolving a racing engine that had first been used in the 1969 European Touring Car series. Unfortunately, the launch of the 2002 Turbo coincided with the oil crisis and the car would only be in production for ten months, with just 1672 examples being made. Porsche meanwhile had already added twin turbos to its legendary 917 Le Mans winning race car – the later 917-30 variant had an astonishing 1100bhp, making it the most powerful circuit race car ever at that point (indeed many rule changes were aimed at ending the car's overpowering dominance). However, it was not until 1974 that the boost technology filtered down to Porsche's road cars. Horsepower limitations had been introduced in various racing series due to dangerously high speeds and cornering, so smaller engines became the norm – it was in this climate that Porsche searched for a way to increase the output of its engines without increasing the capacity. The solution was to fit their iconic flat-six 911 engine with a turbo. Encouraged by the need to homologate the 935 race car into Group 4 competition, Porsche unveiled the 911 Turbo at the 1974 Paris Motor Show.

2M

The 911 Turbo's aggressive styling, such as the striking wide arches and tea-tray rear spoiler, were obvious visual highlights, while the brakes, clutch and transmission all had to be strengthened to cope with the extra power. However, it was the addition of a KKK turbo-charger (the initials stand for Kühnle, Kopp and Kausch) that was the game-changing scientific element (adding around 60bhp compared to the non-turbo-charged Carrera). In retrospect, the advent of the turbo has since become one of the most important technological innovations in the history of supercar science, in ways that could not have been imagined at the time.

"With the first 911 Turbo, speed was definitely back."

This innovative science was far from perfect at this point – the chief drawback was dreadful turbo lag, meaning the accelerator would be pressed but there would be a delay in the power boost arriving … worse still, that delay might be unpredictable in length. Therefore many Turbo owners applied the throttle on a straight, but found it arrived halfway around a bend. Add that to the fact the engine was hanging out the rear and it's easy to see why colossal oversteer was a massive and potentially very

OPPOSITE_The standard and turbo 911 engines, with their respective cars.

dangerous problem. One thing was for certain though, with the first Porsche 911 Turbo, speed was most definitely back.

Without doubt the 911's turbo added a massive power boost, so from that point of view the science had been a success. However, the reality is that the early three-litre 911 Turbos were, and remain, far from easy to drive. Magnus Walker is the world's most prominent modifier of Outlaw Porsche 911s and has a collection of Stuttgart's finest that includes a number of early Turbos, including an ultra-rare 1975 Turbo. He is very well placed to explain the strengths – and very definite weaknesses – of the science behind the early 911 Turbo: "By today's standards, an early three-litre turbo is not necessarily that quick a car. In the modern world, you don't necessarily want to drag race anything from a stop light because you might be a little bit humbled when an average, everyday sports car can dust you off, but that's really not the point of the Turbo. Yes, it has the iconic shape, but the real thrill of the Turbo is it's a challenge to drive. For me it is completely different to any normally aspirated car that I have. The real challenge is the tall gear ratios. Everyone talks about turbo lag, which I'll come to in a second, but the real challenge with the Turbo is the tall gear ratio. Under 3,000rpm it's nothing, it starts to build but then it comes on with a wallop around 3,500rpm, 4,000rpm. First gear generally is 45/50mph, second gear is like 85/90mph, third gear is probably like 120 and fourth gear is essentially the rest. So you're pretty much just in second gear most of the time.

RIGHT_Every detail of a Porsche is carefully considered – inside and out.

TURBOS

Turbochargers perform the same job as a supercharger: both draw in ambient air, compress it and then force the air into the engine to create a bigger explosion and therefore more power. But while superchargers are air compressors driven mechanically by the rotating engine, turbochargers are driven by exhaust gases flowing through a turbine, which drives a compressor that in turn forces in the air.

For turbochargers to work most effectively, some fundamental requirements have to be met: energy, temperature, power, pressure, load, inertia and friction all need to be tamed and controlled. Cooling the turbocharged (compressed) air is vital: the cooler the air entering the cylinders, the better the behaviour immediately prior to combustion.

Early turbochargers such as on the Porsche 911-930 behaved as either "on" or "off", with huge issues of lag – a noticeable delay between putting the accelerator down and the moment when the turbo kicks in. This lag is caused by the compressor taking time to build up air pressure (spooling) and delivering the forced induction.

The Porsche 959 (see page 99) featured twin turbos, each of which had a different power delivery purpose and level. One remained dormant at low speed, enabling slow driving with ease, and better transition under acceleration to full power, with the second turbo coming in when called for. A similar principle is used in the highly advanced two-stage quadruple turbos of the Bugatti Chiron.

Early turbo petrol engines had a low compression ratio – this means the squeeze and bang part of the four-stroke cycle has a larger volume of air but lower compression ahead of the explosion – to compensate for the pressure and heat brought by these early turbos. These early turbo cars were less responsive at low revs. Fuel development and intercooling alleviated this greatly (with compressed air passing through a radiator prior to induction into cylinders), thus allowing the compression ratio of the engine to be increased to a level which was more sympathetic to the turbo-charged boost. A higher compression ratio would produce a smaller volume of air to be compressed but at increased pressure, thereby making that process more efficient.

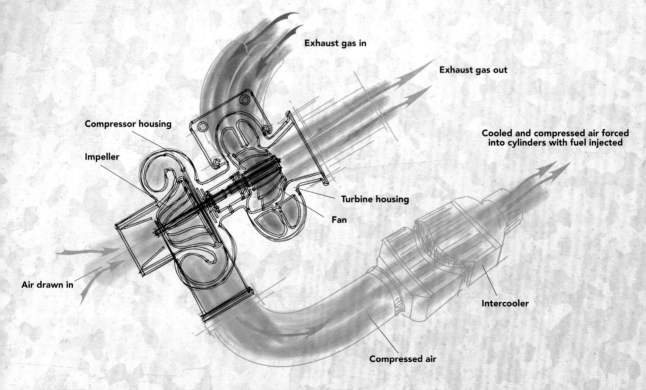

Exhaust gas in

Exhaust gas out

Compressor housing

Impeller

Cooled and compressed air forced into cylinders with fuel injected

Turbine housing

Fan

Air drawn in

Intercooler

Compressed air

"The Turbo is a challenge to drive, you always have to be thinking so far ahead because you can get caught off-guard pretty easily. You can be cruising in third gear on the freeway at 60mph, put your foot down at 3,000rpm in third gear and there's really no acceleration. What you have to do is remember to downshift to second gear, which is really counter-intuitive. The challenge really is to make the most out of the gears and keep your foot planted. I think you need to reinterpret your brain to the gear ratio.

"As for the turbo lag, well, it can be a while. When the turbo finally lights up, you end up eating road really, really quickly, so if there's a car 100 yards ahead when that boost comes on, you're up on the bumper of that car really, really

ABOVE_The 930-911 Turbo lends itself to subtle modification – the model seen here sits low and on bespoke Fifteen52 alloys.

quickly. If you are weaving in and out of traffic, there's not a lot of instant throttle response and acceleration if you're not on boost. So the challenge really comes to what I call boost management, anticipation of when boost is going to spool up and also gear ratio management. The early Turbo is not that refined, really, it's not that subtle, it's kind of like a sledgehammer, it's on or off, there is no in-between. But that's the charm of the old car, right? You can go really, really fast in a modern supercar, with not necessarily a lot of driving ability or experience. An early 911 Turbo is not nearly the same because once you get in over your head, that car is not forgiving – hence the name, the widow maker."

Writing some 40 years after the introduction of the first Porsche 911 Turbo, there is a sweet irony about the supercar's history of forced induction. When the turbo was initially introduced, the aim was to coax even more power out of already high-performance engines. As the seventies gave way to the decade of conspicuous consumption that was the eighties, the Porsche Turbo came to represent Yuppie culture, the drivers who wanted "more bang for the buck" by showing off their success as loudly as possible. However, fast-forward over three more decades and now the turbo has been subverted to become the piece of science that makes cars more economical, emissions compliant and socially acceptable. Turbo technology filtered down into more everyday cars when the science was adapted

ABOVE AND OPPOSITE_The Porsche 911-930 Turbo is a famously savage car, requiring considerable skill on the part of the driver.

into family saloons such as the Mercedes-Benz 300SD and the VW Golf Turbo Diesel in the early eighties. In the present day, millions of road cars have now been fitted with turbos to make their smaller capacity engines meet ever-stricter regulations, rather than to go faster (this is despite the fact that early turbos were actually *less* economical than their naturally aspirated counterparts). By the 2010s, supercar manufacturers began to release turbo-driven cars that boasted smaller engines, were still brutally fast but also capable of meeting stern regulations. Take the naturally aspirated Ferrari 458 being followed by the equally beautiful twin-turbo 488. Not every marque is a turbo convert – at the time of writing, for example, Lamborghini has not yet launched a turbo-engined supercar.

> **"Now the turbo has been subverted: it makes cars more economical and emissions compliant."**

It could be argued that the Porsche 911 Turbo has bequeathed the automotive world with a legacy of economy and carbon sense, perhaps the ultimate irony for a savage performance car. Using exhaust gases to boost power has since proven to be absolutely central to the history of fast cars – in a sense, the science behind the turbo first seen on the 1975 Porsche 911-930 might just have saved the supercar.

CHAPTER 6
GETTING A GRIP

After the relatively fallow years of the seventies, the eighties proved to be an extremely productive and innovative period of supercar science.

The most striking advancements were the use of four-wheel drive to establish more grip, the introduction of lighter materials such as composites, the awareness that a supercar had to (at least) attempt to offer a complete package rather than just one or two scientific highlights and also a glorious return to an out-and-out top speed race.

While tyre technology had been improving in incremental steps, along with suspension and brake evolution, the simple fact was that engines were now so brilliantly engineered and immensely powerful that it was becoming a major problem transferring those forces safely down onto the road. As brake technology often lagged behind, many proud owners met an untimely demise when the impressive straight-line speed of so many well-known supercars was not matched by nimble and grippy road handling. The large number of crumpled supercars dragged out of hedges was ample anecdotal evidence that not all elements of these high-performance cars were created equal. Curiously, however, it took a vehicle that was not ostensibly a supercar to provide the answer – the Audi Quattro.

RIGHT_Porsche 959, 1986

The Quattro marks such a significant moment in high performance car history and is such a capable creation that despite it being essentially a rally or sports car, it deserves to be highlighted in these pages. Just as the Miura was the Year Zero for mid-engined layouts, then so is the Audi Quattro the dawn of the modern supercar's four-wheel drive systems.

There is actually a significant history of 4WD being experimented with: traction engines as far back as the late 1800s used the notion; Ferdinand Porsche (that man again) designed a four-wheel drive electric vehicle in 1899; Mercedes started building 4WD vehicles in 1903 and a Spyker of the same year also toyed with the science. There are many other examples throughout the coming decades, including in military and industrial applications. The Willys CJ-2A jeep, for example, was a famous military-influenced vehicle credited by many as being the first full-production four-wheel drive to be made commercially available. The first road-going production sports car with four-wheel drive was actually the Jensen Interceptor FF from 1966. This car used the namesake Formula Ferguson system that was the brainchild of the wealthy tractor manufacturer, Harry Ferguson.

The vogue for luxury 4WD cars led by marques such as Range Rover was also a notable step along the evolutionary path. Even so, by 1980, it was only really the Jensen Interceptor FF that had

"The Quattro is the dawn of the modern supercar's four-wheel drive systems."

truly attempted to use 4WD in a high-performance car. Eighty years after its inception, no major manufacturers had taken to the science enough to launch a mass-market 4WD car for the road. So it is to the high-speed corners and exhilarating off-road races of the World Rally Championship that historians must look for the most significant championing of four-wheel drive.

Prior to the Audi Quattro, four-wheel drive was banned in the World Rally Championship. Audi questioned this, knowing that they had technology to shock the world up their sleeve. Since 1976 they had been working on a VW-badged car with four-wheel drive called the Iltis (German for polecat), powered by the 75bhp engine from an Audi 80. The brilliant engineer Jörg Bensinger (who would become known as Mr Quattro) conceived the notion and with his team started to build a concept for a car utilizing this system, codenamed EA262. The aim was to showcase the technology in a world championship rally programme, in order to convince the car buyer that 4WD would work well on road vehicles. Legend has it that the revered rally driver Hannu Mikkola was initially unconvinced when he was asked to join Audi's rally team, but after 30 minutes in the (plain-looking) Quattro, he agreed immediately. The road car came first, at the Geneva Motor Show in 1980, with *Motorsport* magazine hailing it as "probably the most significant new road car of the decade". The Quattro made its rally debut at Austria's Janner race in January 1981 and won – by over 20 minutes. World Rally titles followed and the history books were changed forever.

> ## "Audi knew that they had technology to shock the world up their sleeve."

BELOW_The prodigious grip offered by the Quattro's four-wheel drive system slayed the rally competition and made previous road-holding science obsolete overnight.

THE DIFFERENCE BETWEEN 4WD AND AWD

Four-wheel drive is the older, more traditional and off-road derived system. Generally, the power split front to rear is 50/50. Four-wheel drive can be disengaged so that just two-wheel drive is used.

The basic principle of all-wheel drive (AWD) is to deliver to each wheel the optimum level of torque, giving best possible drive and traction. AWD tends to be permanently engaged, but the power split front to rear along with individual wheels is variable,

according to what each wheel is sensing and therefore demanding.

The Audi Quattro featured a so-called Torsen differential (short for Torque Sensing) – a revolutionary means of sensing what each wheel was doing that featured some clever gear mechanisms. Fast-forward to the Bugatti Chiron and a Haldex system is used that incorporates an electro-hydraulic clutch actuator and centrifugal valve, which calls for the delivery of torque transfer to each of the wheels according to demand.

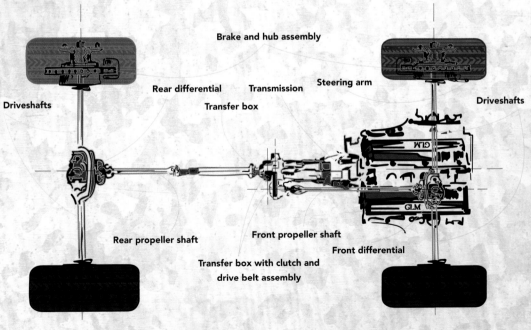

Brake and hub assembly

Rear differential   **Transmission**   **Steering arm**

Transfer box

Driveshafts   **Driveshafts**

Rear propeller shaft

Front propeller shaft

Transfer box with clutch and drive belt assembly   **Front differential**

Sensors send precise values of wheel speed, torque, wheel slip/spin, brake pressure and steering input to the AWD control system, which adjusts the torque values supplied. The transmission drives the propeller shaft to the rear wheels and back through a transfer box driven by toothed belts and/or gears to the front propeller shaft. Clutch plates in the transfer box ensure that torque delivered to front and rear is smooth.

ABOVE AND
OPPOSITE_The Ford
Sierra Cosworth
and Peugeot
205GTi were
both "everyday"
cars that offered
scintillating
performance.

Computational Fluid Dynamics (CFD) pioneer Nick Wirth of Wirth Research has little doubt about the historical importance of this car in terms of high-performance science (and its filter-down technology for everyday cars): "Up to the arrival of the Audi Quattro, 4WD vehicles tended to be Jeeps or Land Rovers. Useful if you wanted to climb a mountain. The Audi Quattro allowed you to use a lot of performance, showcased at a time when World Rally and Rallycross was being widely televised. The DNA from these machines, coupled with the rally results and huge interest, made the perfect storm by giving the road car user close to supercar results. Astounding performance."

"'Ordinary' cars were now boasting supercar-baiting performance figures."

So powerful was the science behind this performance innovation that it was inevitable for designers in the world of supercars to look to utilize what this famous Audi had pioneered. Some marques were relatively slow to the game, partly because inserting a heavy 4WD system into a lithe and light supercar seemed to be self-defeating. It was not until the Audi takeover of Lamborghini in 1998 that the system became more commonplace in the supercar world. The first Lamborghinis to boast an Italian take on 4WD were the later versions of the Diablo; in the modern era, the 4WD system is very popular in supercars, although not universal. For example, without the Audi Quattro, it would seem impossible that the record-breaking Bugatti Veyron of 2005 would have been able to hit 267mph without its tornado-like torque and power sending the car out of control.

It was not just the science in the Quattro that woke supercars up to more fierce and innovative competition – in the wider car world, "ordinary" cars were now boasting genuinely supercar-baiting performance figures, including eighties legends such as the Sierra Cosworth and Peugeot 205 GTI 1.9. Supercars in the eighties had to somehow find the science to "up their game", which, fortunately, is exactly what they did.

This truly was the age of the modern supercar and the science that produced that vast evolution is breath-taking. If the seventies had represented a fight for the very survival of the supercar, the eighties and onwards saw an incredible explosion of scientific advances within the genre that exponentially accelerated over the coming years. The arrival of computer and electronic technology began to have an immediate impact on supercar science, while the power of the turbo charger once again proved its worth (in terms of motor racing, after being introduced into Formula 1 in 1977, the turbo was eventually outlawed for a variety of reasons, a move that in itself saw a swing back towards the redevelopment of the naturally aspirated engine).

While marques such as Aston Martin and Lotus struggled, Ferrari bounced back from the shock of being trounced by the Lamborghini Miura and Countach by producing two cars that almost single-handedly redefined what a supercar was expected to do and, specifically, how the science behind these cars achieved their goals: the Ferrari 288 GTO and its more famous sibling, the F40.

"The savage 288 GTO was
not for the faint-hearted."

The 288 GTO arrived first, in 1984. Superficially taking its design cues from the pretty 308 GTB Speciale of 1977, the 288 GTO was a very different beast. Initially designed to compete in the savage Group B rally series, its entire existence was questioned when that race series collapsed. Undeterred, Ferrari continued developing what many consider to be the world's first hypercar. Why is the 288 GTO acclaimed in this way?

This brutal new Ferrari certainly had several elements of scientific innovation that impressed: twin turbos (the first time in a road Ferrari); electronic ignition and injection systems culled from their F1 team; a ferocious all-alloy V8 engine mounted longitudinally, with the gearbox in line, too. The aggressive body, wide arches, 250 GTO-inspired rear spoiler and air slats behind the rear wheels were all the height of supercar style. The 288 GTO was a savage, brutal beast complete with a blood-curdling howl of a soundtrack when the turbos kicked in. Not for the faint hearted. Impressive stuff, but not necessarily a quantum leap.

However, it was the science behind the GTO's construction that was a revolution in the supercar world. Car designers had from day one been searching for ever lighter vehicles, best summed up by Colin Chapman's famous obsession with his desire to "Simplify, then add lightness." Until now the science of materials including aluminium, alloy and fibreglass had not pushed the envelope enough. More weight-saving measures were needed.

So how exactly did Ferrari do that with the 288 GTO? By using so-called composite materials in its construction, such as carbon fibre that had previously been the preserve of racing cars (along with other immensely strong synthetic fibres such as Kevlar).

The 288 GTO plundered these ultra-strong, very lightweight racing materials in its design. For example, the designers used Kevlar in the tail and rear bulkhead, moulded fibreglass and carbon fibre in the roof and aluminium honeycomb in the firewall. The end result of this pioneering weight-saving construction was a car that was 113kg (250lb) lighter than the already lightweight 308GTB Speciale. F1 design guru Adrian Newey recognizes the significance of this composite science: "My first job in motor racing was at Fittipaldi, where the technical director was Harvey Postlethwaite who then left for Ferrari. Among the things he is quite proud of is the work that he had done on the 288s, particularly the honeycombed composite bonnets and some of the other lightweight bits put on the car. I subsequently bought a 288 GTO myself."

In terms of scientific firsts, the 288 GTO marks an important moment. After the 288 GTO, if a supercar was to be *really* fast, it had to use composites, a science that within a few years became de rigeur for any self-respecting high-performance machine.

As science progressed within the supercar world, some manufacturers adopted the manifold benefits more than others, not necessarily out of a greater vision, but from a different philosophical standpoint. Perhaps no two cars typify this more than the famed rivalry between Porsche's engineering tour de force, the 959, and Ferrari's famously brutal F40.

Porsche had been at the forefront of the genre almost since the advent of the motor car and the Stuttgart factory's response to the Ferrari 288 GTO was a staggering exercise in scientific brilliance: the Porsche 959. For many observers, the 959 is *the* most scientifically advanced car of its generation and arguably a supercar that raised the stakes in terms of what consumers expected of the genre. The 959's elongated looks polarize many, and it is certainly not as striking as the iconic Ferrari F40, but the science behind this particular supercar was remarkable.

Like the Ferrari 288 GTO, the Porsche 959 had been designed with Group B rallying in mind, a series which stated that two hundred road-going examples needed to be built to qualify – in return the competition prescribed *no limit* on horsepower. When the mid-engine revolution started by the Miura had taken hold (even Porsche themselves had introduced the mid-engined 914), the days of the rear-engined supercar seemed numbered. Porsche did toy with ending its 911 range, but by the mid-eighties decided instead to see how far they could push that concept. In 1981, a one-off all-wheel drive 911 was shown at Frankfurt (the same year that the Quattro entered rallying). Two years later a similar pioneering car won the gruelling Paris–Dakar rally, quickly followed by the

"For many, the 959 is the most scientifically advanced car of its generation."

959 concept car, known as the Gruppe B. The 959 itself was launched in 1985 but the technological demands saw deliveries not commence until 1987 (after Group B had been abandoned). Yes, the 959 was the fastest car in the world at that time but it was also the most scientifically advanced supercar ever made. The list of technological innovations is long and impressive. The four-wheel drive was computer controlled, making the 959 arguably the first true supercar to use that technology. It offered four settings: dry, wet, snow and off-road. Significantly, the car boasted so-called torque vectoring, which controlled the way the drive from the twin turbo to the rear-mounted engine was distributed to each corner, depending on the driver's demands and the conditions.

"The 959 offered four settings: dry, wet, snow and off-road."

BELOW_Such was the breadth of the Porsche 959's performance envelope that it was able to compete in the Paris–Dakar rally.

TORQUE VECTORING

Torque vectoring enables the driven wheels (and therefore tyres) to supply maximum traction independently of each other. All-wheel drive supercars make full use of torque vectoring. Electronics, sensors and controls make this possible as feedback from what each tyre is "feeling" is essential – effectively attempting to grab and grip the road surface, thereby increasing traction. This is then reported back to the supercar's brain (the Engine Control Unit or ECU), through sensors, providing the requirements from each corner of the car, such as information on rotational speed, steering angle, then creating a demand (from the engine) for torque from the driveshafts transmitted through the wheels. The required torque is fed to each wheel via clutches working in conjunction with the differential (and driveshafts), so friction is another force at work here. Torque vectoring has especially come to the fore with recent supercars and is still being developed and understood. VW/Audi have been deploying torque vectoring since the eighties (the Quattro etc.) and transferred this tech over to their saloon cars, but in supercar terms the science was first displayed in the Porsche 959.

Electric motors

Reduction gears and driveshafts

Torque vector clutches

Power supply/invertor

Motors braking — recharging batteries

Motors delivering variable torque

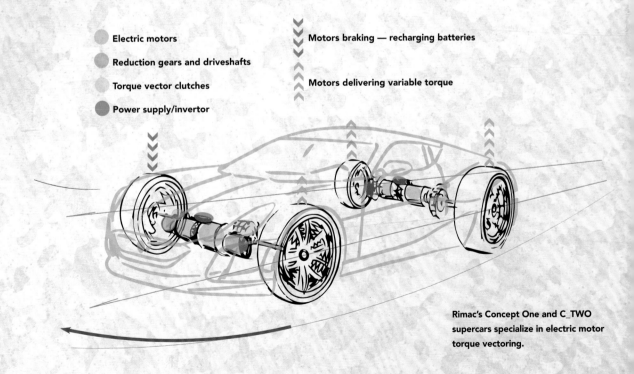

Rimac's Concept One and C_TWO supercars specialize in electric motor torque vectoring.

The 959's ABS brakes were also computer controlled, the steering had servo-assist and the ride height and suspension stiffness were variable. Other meticulous scientific details included tyre pressure monitors and lightweight magnesium alloy wheels. Even the turbos were evolved – as mentioned earlier, the KKK units were sequential, meaning they were designed to be more gradual and smooth with the delivery of boost, rather than the more scary option of "all at once". Drag coefficient was minimal while downforce was mightily impressive, although critics did not take to the long-tailed looks that achieved this twin accolade. With the plush interior, modest cabin noise, a soft ride, relatively soft clutch and excellent visibility, the 959 had used every ounce of known supercar science to create a new landmark in driveable supercars. From a scientific point of view and judging the car in the context of its time, Porsche had moved the goalposts for supercar manufacturers. In the years after the 959, it was no longer acceptable to release a supercar that was not a coherent, well-planned and scientifically advanced vehicle.

The Porsche 959 was an exercise in pure unadulterated science, but representing a polar opposite to that philosophy is the iconic Ferrari F40, which was launched in 1987 to celebrate the 40th anniversary of that famous brand. Its scientific approach

RIGHT AND OPPOSITE_For many, the Ferrari F40's silhouette remains perhaps the most famous supercar design ever created.

to high performance could not have been more different. Here was a stripped back, raw and brutal example of a race car for the road. Ferrari achieved this without a monstrous V12 engine, instead using a twin-turbo 2.9-litre V8 – an important moment in supercar science in itself. Its lightweight construction and racing heritage meant this was the very first road-going car to hit 200mph, a massive scientific moment for the genre.

"NACA ducts reveal the F40's racing genesis."

Drawing tech directly from its racing team, Ferrari expanded upon the manifesto of the 288 GTO by using even more composite materials such as Kevlar, carbon fibre and aluminium (with many joints glued to save weight). NACA ducts punched into the bonnet and scoops in the side reveal the F40's racing genesis. A plexi-glass engine cover saved weight, as did a lack of carpets or even door handles (a piece of cable was used). Composite was used for the 1.5kg (3.3lb) seats, while electric windows were not available (too heavy). No ABS, no power steering, no brake servo, no airbags, no central locking … Ferrari went even further, using fewer coats of paint on the beautiful silhouette to save yet more weight. Yet the marque was not afraid to utilize F1 tech where needed – the F40 has innovative injection and ignition management systems directly influenced by their F1 programme.

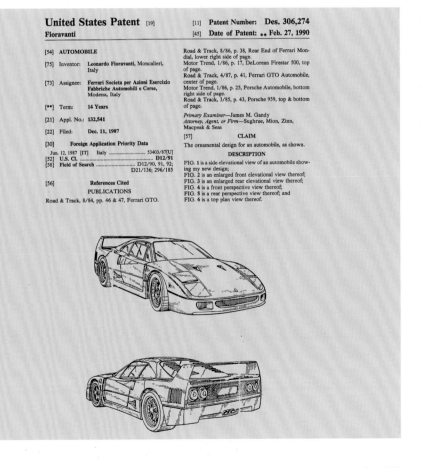

NACA DUCTS

NACA (National Advisory Committee for Aeronautics) ducts are a simple means of taking fast-moving surface air into a vehicle for cooling. NACA ducts originated in US aeronautics and are effectively a low drag air scoop or intake. NACA ducts were incorporated in the design of the Ferrari F40 to help solve the cooling challenges created by the hugely powerful twin turbo 2.9L V8.

Conventional ducts are so-called Ram ducts. These face the air directly perpendicular to the motion of travel. Although this approach allows a large volume of cooling air in to the duct, it comes with a penalty of aerodynamic drag.

By contrast, a NACA duct is set into the bodywork of the car, effectively below the surface of the panels, which allows it to draw cooling air into the duct as the car travels along. The airflow over this in-set surface is subject to a suction force because air accelerates into the NACA duct as the scoop becomes wider, thereby supplying much-needed cooling air.

A significant challenge when using NACA ducts is to ensure that the ducting beneath or behind such ducts is designed effectively and also sealed correctly, to avoid stifling the cooling air that has been generated. In badly designed supercars, poor ducting design can render NACA ducts almost useless.

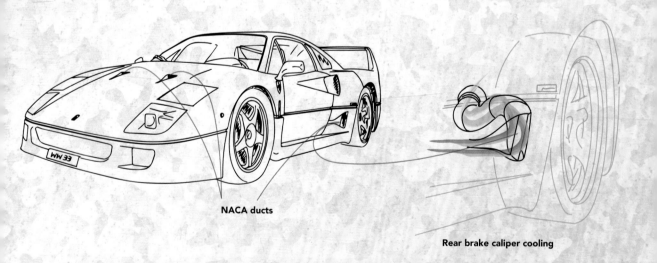

NACA ducts

Rear brake caliper cooling

The F40 is regarded by many as the greatest supercar of all time. Its beautiful exterior styling certainly created one of the all-time great supercar silhouettes. The F40 was a huge commercial success, but for the purposes of this book, it is the raw, stripped-back approach to science that makes it so important. This was effectively Ferrari saying, "Just because the science is available doesn't mean it has to be used to excess." Aesthetically, this was also the Italian marque clearly stating that art remained as important as science in the creation of brilliant supercars. The

"Art remained as important as science in the creation of brilliant supercars."

fact that in the modern day the F40 is almost universally regarded as a more important high-performance car than its rival the 959 (or certainly a more historically desireable one) is tribute enough. The rivalry between these two cars in the mid-to-late eighties perfectly sums up how the science of supercars was exploding almost week by week, a rapid acceleration of tech that in itself presented many new challenges. How designers, engineers, manufacturers and, of course, supercar customers reacted to these very different solutions is very telling.

Interestingly, legendary motorsport designer Adrian Newey is not a big fan of the F40: "A friend had one and I must admit that while it's in the category of creating a good driving experience, arguably better than the 288 GTO, although somewhat less civilized, to me the build quality of the F40 was shocking. And personally, I hate the styling. I know some people love it, but it's not to my taste at all."

But it is this very styling that appeals to many, including those in search of a little more excitement and drama. People such as musician and supercar fanatic Liam Howlett of The Prodigy, who is a huge fan of the F40. His attitude to supercars reflects the opposite approaches that Ferrari and Porsche were taking at this point: "With any car I've owned, just being fast is never enough. Handling goes without saying, but it's got to have drama too, it has to make you feel excited just sitting in it. The F40 has that."

The explosion of supercar science in the eighties created a fascinating dichotomy. As the technology rapidly improved, the role of the driver was diluted. The ultimate manifestation of automotive technology will be a fully self-driving car; however, to a supercar fan or customer, that is when the genre is ruined. How supercars adopted technology over the years while trying to maintain the sensory experience of being in control at the wheel is a key element in the story of supercar science. Some threw every gadget at a car as the ultimate next step; others deliberately pared back the science so that the driver felt more in control. Whatever the philosophical approach, the fact is that in terms of supercar science, by the end of the eighties the genie was well and truly out of the bottle.

"Just being fast is never enough."

Liam Howlett, musician

ABOVE AND OPPOSITE_ The F40 was a genuine "race car for the road", although some racing examples were tuned to be even more powerful. So raw was the lightweight standard spec that even seasoned supercar drivers could be caught out by the car's savagery.

CHAPTER 7
THE COMPLETE PACKAGE

Throughout much of the seventies, and to a lesser degree in the eighties, the world had been remarkably tolerant of supercars.

Very difficult to drive? No problem. Zero visibility? Minor issue. Burn the clutch on a slight hill start and spend a small fortune to replace it? To be expected. Awful ergonomics, no space for luggage … all fine. The list went on. The many foibles and faults of supercars were largely tolerated, perhaps even enjoyed, as part of the ritual of owning one. To possess such an extravagant beast almost demanded that ownership was problematic. Clearly some cars excelled, such as the powerful F40 and 288 GTO, the tech ambition of the 959, and so on, but there was also an element of "either/or" with the high-performance cars on offer at the time – either luxury or brutality, with few cars competently offering both.

By the late eighties and early nineties, this state of affairs was no longer acceptable. Family cars at a budget price were becoming crammed with technology that shamed many high-performance machines. The wealthy supercar owner began to expect more, and rightly so. The next quantum leap for supercars had to be a machine that offered the *whole package* – both in terms of what it was capable of but also the way that science was physically designed and delivered within the car.

The question for supercar designers was this: did science allow the creation of a car that was driveable, practical, reliable,

RIGHT_McLaren F1, 1992

enjoyable, yet still had blistering, other-worldly performance and, of course, stunning looks? Moreover, in the early nineties, severe recession had blighted the consumer market for luxury goods across the globe and pretty much seen the supercar market collapse. So was there even enough money – and demand – to develop, launch and then ultimately sell such a perfect machine?

The answer was yes and it came in the form of Gordon Murray and his acclaimed McLaren F1. Launched in 1992 by a prestigious racing team that had no previous history of selling a series production road car, the McLaren F1 shattered every performance benchmark and was so far superior to its rivals that it was almost humiliating. If the F40 and Porsche 959 had previously been engaged in a supercar race to the top, then the F1 was in a contest entirely of its own.

South African-born Murray is famous for his countless Formula 1 achievements and legendary race car designs (including the revolutionary Brabham fan car), but his history in terms of car design goes much further back: "When I was growing up in Durban in the fifties, the only racing cars were really handmade specials. People didn't really have much money after the war to buy or race cars, so they built these specials. I was a surfer, I was into bikes and cars, racing and, of course, rock and roll in equal measures. George Lucas's first film, *American Graffiti*, is my upbringing. That's what I did, drive-in movies, milkshake bars, all these hot cars every Sunday night, bonnets up. I had a '56 Hillman Minx because that was all my dad could afford, I took the hub caps off and painted the wheels silver and added a noisy exhaust and I used to drive around with the choke pulled out so it sounded like it was tuned.

"We all used to tune Fords in those days, Anglias and Cortinas with 1640 engines in them and Weber carbs, and they would be pretty quick little cars. I can remember seeing an AC Ace for the first time and thinking, *Bloody hell, that's quick!* That is my first recollection of anything exotic. Somebody else had an E-Type. The first exotics I saw after that, funnily enough, were road/race cars. The ones that really caught my eye in the sixties were things like the 904 Porsche, 275 LM Ferraris, 250 GTO Ferraris, Ford GTs, all of these wonderful road/race exotics."

Art classes and a background in technical drawing give some hints of the future blend of art and science that Gordon would be able to bring to the McLaren F1. A love for Colin Chapman's Lotuses and the pursuit of lightness, as well as speed, further underpinned Murray's driving force: *an obsession with detail*. "For three years at Brabham, I was on my own in the office. So I did gearbox design, engine design, chassis, aerodynamics, suspension geometry, fuel systems, cooling systems on my own," Murray explains. Within this quote lies the genesis of the F1's comprehensive and sublime packaging.

> ## "I thought, *Britain should be making these things. Great-looking things, great-sounding things, very fast.*"
>
> Gordon Murray, designer of the McLaren F1

Remarkably, when McLaren announced they were going to build a supercar with Murray at the helm, he had yet to work on such a project before, but he was openly dissatisfied with the supercar science that was on offer.

"Every time I borrowed a Lamborghini Countach, or a this or a that, or saw one close up and looked underneath to see glue and cardboard or something – these cars could be so disappointing. I happened to look underneath a Ferrari Dino 246 and it was just horrible, such badly welded tubes and an awful frame," Murray says. "I always thought, *Britain should be making these things. Great-looking things, great-sounding things, very fast.* I had probably subconsciously started building a library of bad things and when I finally got the chance to do the F1 on a clean sheet of paper I was determined to exorcise all the elements you shouldn't do – uncomfortable pedal offsets, bad visibility, no luggage space, air-con that doesn't work, a sound system that sounds like it was chucked in at the last five minutes. Not good enough."

ABOVE_Niki Lauda driving a Gordon Murray-designed Brabham BT46 fan car in the 1978 Swedish Grand Prix at the Scandinavian Raceway in Anderstorp, Sweden.

PACKAGING

Packaging is a vital part of any supercar's layout, architecture and design – basically, where do all the assemblies and sub-assemblies fit within the car envelope (including the occupants) and are they optimized from the perspective of car performance, safety and ergonomics?

Supercar architecture and specification constantly adapts to accommodate the latest innovations adopted by the manufacturers. The arrival of energy recovery/hybrid technology with cars such as the Holy Trinity (see page 174) brought with it new packaging challenges. The powertrain and drivetrain (engine and gearbox) for supercars up to the arrival of the Porsche 918 was common across all marques and models. For 100 years, headline ingredients required to create a car and supercar were:

- Internal combustion engine (ICE)
- Exhaust system
- Fuel system
- Transmission and hydraulics (and/or pneumatics)
- Suspension, brakes, wheels and tyres
- Driver controls and interior (seats, safety)
- Chassis
- Bodywork
- Cooling system
- Electrics and controls

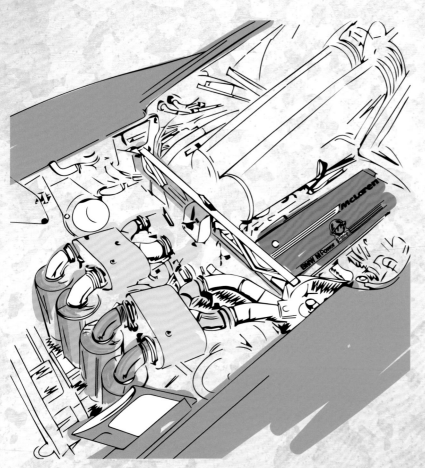

Such was Murray's strength of vision about the packaging and presentation of the *entire* car that the development process began with a day-long briefing. "When I started out, I could picture the whole car in my head so I had an infamous ten-and-a-half-hour meeting where I got the team together. Once I had picked everybody, a lot of them were youngsters, I locked the room and nobody got up for anything for ten and half hours while I told everybody what the car would be, what it would represent, how the customer would be treated, how the customer would feel about the car. What I was trying to achieve was much more than car design on the F1. I wanted to set completely new standards for customer experience, the way you treated the owners, even bespoke seating … building the car around the person so they felt incredibly special. I know a lot of that stuff is commonplace now but I believe when we did that it was the first time."

The performance elements of the F1 were staggering, using state-of-the-art technology. The custom-built 6.1-litre BMW Motorsport V12 naturally aspirated engine produced 627bhp (559bhp per tonne) and was the most powerful engine of its kind used in a road car until the Ferrari 812 Superfast was launched some 26 years later. It was fed oxygen by an aggressive-looking, large rooftop air intake. The transverse gearbox helped keep the overall length of the F1's powertrain very modest, hence the relatively small footprint of the car.

Race car materials such as magnesium (for example, the race-style alloy wheels), aluminium and carbon fibre were used extensively and without limitation. If the performance and packaging benefited then the very best was signed off,

ABOVE_ The even more ludicrously fast F1 GTR.

ABOVE_The F1 was famous for its three-seat layout, with a central driving position.

OPPOSITE_Gordon Murray was absolutely obsessive in his attention to detail and in his vision to create the greatest driver's car ever made in the McLaren F1.

including gold leaf being used as the best reflector of heat for the underside of the engine lid. A three-plate carbon-on-carbon clutch, a complex exhaust system, a racing-inspired fuel cell and so-called intelligent brake cooling were other scientific elements of note. To assist the massive disc brakes, there was a pop-up spoiler for high-speed braking, which was considerably ahead of its time on road cars.

The all-carbon fibre bodywork designed with Peter Stevens (who had worked on the Lotus Elan, Esprit and XJR-15 Jaguar) was beautiful, complete with spectacular so-called dihedral doors. Despite its long wheelbase, the F1 was strikingly compact, yet another nod to the genius of the packaging. Weight distribution was honed to avoid the so-called dumb-bell effect, with all the masses contained within the wheelbase. The mid-engined, rear-wheel driven featherweight car had minimal rear and front overhang and a low centre of gravity; this sat on a carbon fibre chassis with a carbon fibre monocoque drawn straight from Formula 1 science, reinforced with aluminium honeycomb panels and Kevlar. Each body took 3,500 man hours to build.

"I stole Formula 1 technology to build the F1. That car was the first carbon honeycomb supercar. There are different ways of using carbon composites – the F1 uses structural composites. A lot of the modern supercars use just a monolith [solid carbon], which is not the best method. Essentially you are replacing a sheet metal skin with a carbon skin and, yes, you get a decrease in weight and an increase in stiffness, but it's marginal. However, if you use Formula 1 technology, you have two thin skins of carbon with honeycomb in the middle, which is what John Barnard started [he created a carbon fibre composite chassis for the McLaren F1 team in 1981]. I used monolithic carbon in Formula 1 cars three years before Barnard but then he took the big jump and did the honeycomb thing. You get multiples better – efficiency from crash stiffness, and weight saving on a huge scale."

> ## "As a driver's car, the F1 still holds its own."
>
> Gordon Murray, designer of the McLaren F1

Due to the composite materials and construction methods used in the F1, the car was ultra-light, weighing in with a dry weight of just 1090kg (2,403lb), slightly over their initial goal of below 1,000kg (2,204lb) – but several hundred kilogrammes lighter than rivals such as the Jaguar XJ220 and Ferrari Testarossa. Obviously the Formula 1 team at McLaren assisted the project greatly with the very latest composite technology; however, during his time at Brabham, Gordon had long been using composite materials from aviation and aeronautical science.

Uniquely, the F1 used a central driving position, which eliminated many supercar ergonomic flaws and provided a real-world "race car" experience with excellent visibility. This was aided by beautifully simple yet purposeful instrumentation. The philosophy of creating the perfect driver's car extended to every level of the science, dictating that Murray did not use traction control, anti-lock brakes or even a brake servo as he believed that tech would overwhelm the driving experience. "As a driver's car, the F1 still holds its own against any modern super- or hypercar. A lot of it is due to legislation to be fair, but what's happening is there is so much electronic interference that you virtually have got nothing to do with the car. To avoid that at all costs was my intention. I wasn't being bloody-minded in not having ABS, power steering or power brakes and so on; it was because I wanted that to be the *purest* driving experience." Yet Murray was able to include leather upholstery, air-con, electric windows, custom luggage, a high-end (lightweight) Kenwood CD player and space for luggage. The carbon bucket seat and pedals were all custom-fitted to each customer.

Packaging presents potential challenges and trade-offs with the laws of physics and forces such as optimum weight distribution, centre of gravity and minimal frontal area (for aerodynamic efficiency). These fundamental principles will always try to dictate the layout and architecture of any given car. In addition, the supercar designers are having to accommodate practicalities such as luggage space, cabin room, comfort, looks – the list of demands on their packaging ingenuity is almost endless. To package a high-performance vehicle which is able to spin all of these plates in a stylish and high performance form is essentially the holy grail of supercar design.

ABOVE_The
F1's cabin was
ergonomically
sublime, with a
beautifullly clear
instrument panel.

Murray achieved this purity of science and design without introducing clutter. "If you look at the F1, there is nothing on it that doesn't need to be there. Even though it has got all of the creature comforts, there is nothing doing nothing. There is not some random bracket that doesn't do six jobs – that is a racing car principle, whereas with many of these modern supercars, if you look into the engine bay they're just a mess, I mean, you wouldn't want to lift the lid and show your mates.

"What I didn't realize back then, and this goes back to the subconscious racing car designer mentality, is that I *can't draw* bad suspension, bad chassis, bad engineering. Instinctively, I designed a car that was light, it had very low centre of gravity, it was the world's first ground effect road car so it had all the racing car elements you need such as a stiff chassis."

To expand on Murray's point, the F1's undertray was inspired by Formula 1 tech, to hold the car onto the ground with rear diffusers and two fans to remove boundary air (directly inspired by Murray's 1978 Brabham F1 car). The centre of the nose sent air under the car, where it was manipulated by the tail diffuser to generate downforce. Rather than push the car down onto the road with spoilers, the F1's clever science underneath *sucked* it onto the surface. In summary, the entire philosophy of the F1 as the ultimate driver's experience raised supercar science to previously unheard-of levels.

> **"There is nothing on [the F1] that doesn't need to be there."**
>
> Gordon Murray, designer of the McLaren F1

McLaren launched the F1 in 1992 as the fastest and also most expensive road car ever (at £635,000/$900,000, the most extreme and expensive versions of the car cost £500,000 /$710,000 more than some Ferraris). Inevitably the headlines were about the blistering top speed (242mph), the 0–60mph times (3.2 seconds), the 0–100mph times even (6.3 seconds). The car was also able to go from 30mph to 220mph using only sixth gear. Yet Murray has a surprising view of the universal plaudits that the car won: "I think the practical side of the car was completely overshadowed by the performance. However, top speed was never an issue, it was *never* a target for me. It was always going to be quick and it is still the quickest normally aspirated car on the planet. It was *all* about the driving experience and that was why the driver is sitting in the middle, why there are no pedal offsets, good visibility, a normally aspirated engine, instantaneous response, no fly wheel … but the practicality just gets completely overshadowed. Even with the modern-day supercars, many of them can't fit in any more than a small weekend bag, and you need the battery to be fully charged at all times. That's not practical nor was it acceptable to me. I set out to try and build *the best driver's car there has ever been.*"

In 1998 Le Mans winning racer Andy Wallace took an F1 to 242mph, a speed record that was so far ahead of its rivals as to be almost ludicrous. "Back when we did that record," recalls Wallace, "that speed was absolutely unheard of, 240mph+. The only time I had ever been that fast before in my life was Le Mans,

pre-chicane. I think even with the first F1 run we were already up over 230mph. It was unbelievable. It's incredible to be sitting in a real road car and actually seeing everything fly by so quickly."

It is interesting to hear this reference to Le Mans because Gordon Murray drew partly on his admiration for the Miura for inspiration for the F1, rather than any race car. "That Lamborghini was never intended for, and never went, racing. That is also what I tried to do with the F1. I said to [McLaren chairman] Ron Dennis and [entrepreneur/F1 team executive] Mansour Ojjeh, "Don't tell me we are going to go racing with this car because I am a racing designer, I am then going to compromise the bloody thing. We have all agreed it should be something you can jump in, drive to the south of France in comfort and if you tell me it is going to go racing then I will make the tail much longer, I will be thinking about where you put the wing, it will look ugly and it will be heavier," ... so they said, "No, no, no, there is no way we are going to go racing with it."

Eventually, of course, Murray's road car did go racing (reportedly due to customers keen on a race debut). The F1 became the first road car to win at Le Mans since the forties. Julius Kruta, Bugatti's official historian, believes that what the F1 achieved with its stunning Le Mans victories will be the last time that a modern supercar can successfully bridge that racing heritage: "Gordon has said that many, many times, he never had it in his mind to race that car, the F1 was purely designed as a road car, but it was capable of being raced. This was part of the

"The F1 was a road car that happened to make a good racing car."

Gordon Murray, designer of the McLaren F1

DNA of his supercar. What it achieved was simply incredible." Murray continues: "So the F1 is absolutely the opposite of a 250 GTO Ferrari which was designed as a racing car, but they built road cars to go racing with homologation specials, drawing on lessons learned in series such as the American Can–Am races. The F1 was completely 180 degrees to that ethos, it was a road car that happened to make a good racing car by accident, if you like. But above all, it was the best road car in the world, the pinnacle of a driver's car."

To many experts and fans alike, the F1 still is the pinnacle. What was achieved with the science and design of the F1 was much more than a top speed figure that would stand for 13 years. It was much more than stunning looks, blistering performance and driveability. It was the seamless design of all of that in one car, with a philosophy that accepted no compromises. Of the many supercars that have changed the game with their science, the F1 altered the landscape of what was expected to such a massive degree that, to this day, it is still considered by many to be the greatest supercar of all time and the standard all future supercars should be measured against.

The F1's technical achievements are sufficient to justify a book in its own right (several have been written). However, the real achievement of the F1 is indeed how beautifully packaged the car is, a direct result of Murray's relentless and obsessive quest for a coherent solution. "Packaging makes a good car or a bad car," explains Murray, "regardless of what it looks like and what it goes like. For me, *car design is packaging.*"

CHAPTER 8

A NEW DAWN

In the aftermath of the F1's devastating speed records and universal acclaim, few supercar manufacturers dreamed of competing with Murray's sublime creation.

OPPOSITE_ The use of composites on the Pagani Zonda took the art of the supercar to new levels – all the more remarkable for having been achieved by a newcomer to the world of high-performance vehicles.

However, that legendary car did polarize opinions about where the science of these machines was taking the genre – some marques chose to offer brutal, adrenalin-filled driving; others proudly championed every state-of-the-art scientific advance currently known. However, it was the extent of the F1's use of the science of composites that represented a new dawn for supercars. In the early nineties, the primary structures of supercars were generally still made from materials such as aluminium (Jaguar XJ220), a mixture of carbon panels, aluminium and/or steel (the Bugatti EB110, Ferrari F40) or steel (Porsche 959).

The F1 was not the first supercar to use composite materials – as highlighted, the 288GTO and F40 had experimented with the idea some years before, as had a number of other cars. However, it was the extent and precision to which the F1 used composite materials that changed the scientific game, using their structural strength and lightweight properties to radically alter the weight, performance and robustness of that particular supercar. A perfect specific example of the comprehensive use of composites in Murray's masterpiece that would soon become a staple ingredient of any high-end supercar was the carbon fibre monocoque – a real joy for any supercar designer.

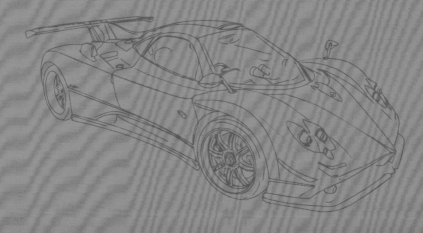

RIGHT_ Pagani Zonda, 1999

THE MONOCOQUE
—— CHASSIS

A monocoque is basically a container. It is comprised of a single coherent unit (when all its components are unified and bonded) as against several separate elements joined together. A monocoque brings great mechanical advantages over early examples of differing types of chassis. Spaceframes, torsion box and ladder frame chassis did not utilize composite technology and as they were usually metal (generally steel) in construction, had strength versus weight limitations. Aluminium monocoque chassis improved this by being light and reasonably strong, but are nowhere near the strength of one made using carbon fibre composites.

The McLaren F1 featured a full composite chassis which offered far superior torsional (twist) and bending stiffness to any other chassis at the time. A fundamental design advantage, this gave a precise and predictable platform from which the systems of the F1 (engine/gearbox/suspension, etc.) could fit. As well as being supremely light, the carbon fibre monocoque also offers huge strength benefits and crash safety improvements.

Composite materials technology and manufacturing techniques enabled McLaren to lead the way. Conventionally, composite materials are expensive and cost-prohibitive for "normal" production cars, largely due to the complex and labour-intensive processes required to not only make the chassis, but also the materials, such as "pre-preg" (pre-pregnated with resin) woven carbon cloth. Tooling (and/or patterns) are also typically very expensive.

Full moulded carbon fibre chassis

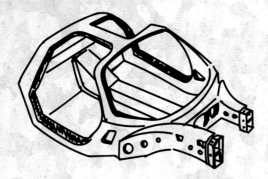

Half "tub" carbon fibre chassis

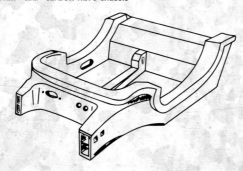

Carbon fibre engine frame

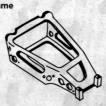

Also known as the tub, the monocoque itself was not a new idea. Most early cars used the more traditional approach of a so-called body-on-frame, which derives the majority of its structural integrity from the frame itself. A monocoque instead uses the shell as a major contributor to structural strength. The idea had been trialled as far back as 1912 but never widely adopted, as the body-on-frame approach was far more well-known, due to its use in the horse and carriage industry. Some historians cite the 1922 Lancia Lambda as the first genuine monocoque in an automobile, because it featured (crucially) a load-bearing unitary body. Although other manufacturers such as Opel and Citroën also experimented with the idea, the concept did not truly take off until decades later in motorsport.

In the fifties, racing cars such as the Jaguar C- and D-Types had used a set-up that was considered years ahead of its time. Legendary F1 designer Adrian Newey is keen to recognize the Jaguars' contribution to the science of the monocoque: "Arguably you could say that the D-Type and then the E-Type, in terms of introducing a proper aluminium monocoque and using aircraft practices in sports cars, were significant. There were other monocoque cars around, but in terms of sports cars, those two Jaguars were important. They certainly paved a new direction long before monocoques were necessarily being used elsewhere in racing." Elsewhere, Colin Chapman's Lotus 25 of 1962 is considered by many to be the first true stressed monocoque racing car, utilizing an aluminium alloy monocoque (around the same period as the Lotus Elite). However, it was the McLaren F1's combination of the

"A monocoque uses the shell as a major contributor to structural strength. "

monocoque theory along with carbon fibre materials that was
the watershed moment for supercar science.

So what exactly are composite materials? Composite
materials had been around for centuries, for example the
Mesopotamians in 3400BC used various strips to form primitive
plywood. Fifteen hundred years later, Mongols crafted bows out
of composites of wood, bamboo and bone. The real startling
progress evolved within modern chemistry, especially early
synthetic resins and plastics, although these materials suffered
from fragility. The thirties was a formative decade: the first
fibreglass was launched, as were early epoxies. These advances
were fast-tracked during World War II and
a fully composite-bodied car was even
prototyped by scientists as early as 1947.
The theories behind this car were later used
in some form for the 1953 Corvette. By the
fifties, the production of strong, flexible and
lightweight fibreglass and other composites
was becoming widespread, especially in the marine industry.
In 1961 carbon fibre was patented and chemical companies
began weaving thin strands of carbon molecules into fabric
to manufacture incredibly strong material. Britain was at the
forefront of this new science, with engineering geniuses at places

> "A fully composite-bodied
> car was prototyped by
> scientists as early as 1947."

such as the Royal Aircraft Establishment and Rolls-Royce Civil Aerospace pioneering the use of carbon fibre composites. The aviation industry moved away from the science after a number of tests saw the formative materials shatter during bird strikes, and for some years the progress stalled.

Then in 1981 the McLaren Formula 1 team revolutionized the construction of racing cars with the first ever carbon fibre monocoque, with designer John Barnard at the helm. He had found little enthusiasm in the UK for his revolutionary idea, but eventually teamed up with an American company, Hercules Aerospace, to create the carbon fibre parts he needed for his race car. Even then, on hearing of his plans, rival F1 teams were sceptical, suggesting that the "black plastic" might just vaporize into dust under a heavy impact.

Undeterred by the doubters and convinced of the science, Barnard's pioneering MP4/1 rolled out of the garage on 5 March 1981 and in the hands of John Watson took a first victory at that year's British Grand Prix. However, it was an appalling crash at a later race in Monza that was the turning point for the science of composites. "Lots of people in the paddock thought I was dead," Watson told the BBC. "I know people who watched on television and they cried." In fact, Watson was unhurt. As McLaren themselves said, the carbon fibre monocoque "enabled

BELOW_McLaren-Ford driver John Watson of Great Britain on his way to winning the 1981 British Formula 1 Grand Prix at Silverstone in a car that revolutionized the sport – by being the first F1 car to use carbon fibre.

ABOVE_The Pagani Huayra uses carbon fibre as a performance aid, but also for its aesthetic qualities.

the Ulsterman to walk away unscathed from a high-speed crash ... over the course of a season it rendered rival chassis obsolete." A new dawn had arrived.

In the aftermath of Watson's crash, Formula 1 teams scrambled to adopt the new carbon fibre technology. In time, supercar manufacturers would also chase the massive potential for this science – the advantages of a strong, lightweight and stiff chassis were invaluable, offering multiple benefits over more traditional methods. The benefit of lightness is obvious – a steel chassis offering the same stiffness as a 100kg (200lb) carbon fibre equivalent one might weigh as much as 1.5 tonnes! Aside from examining technical data and drawings, it is equally striking to watch a famous social media clip that shows a Lamborghini smashing into a tree at speed and literally being scythed in two, yet the monocoque tub – and the driver inside – emerge relatively unscathed.

The obvious downside to creating a supercar with so much carbon fibre is cost, largely due to the inordinately slow process of manufacturing composites – recall how each F1 body took 3,500 man hours to make. However, with affluent supercar buyers happy to pay huge premiums for cars with this science, cost was less of an issue. Composite technology was here to stay.

CARBON FIBRE

A single strand of carbon fibre is approximately five microns, about fifteen times thinner than a human hair (1000 microns = 1mm). These fibres are made from carbon atoms and crystals. The fibres are grouped as a form of thread and this thread is then used to weave cloth, which is then coated with resin as a bonding and curing medium. Each ply of carbon fibre cloth is placed in a mould in a particular orientation (in relation to the other plys) to maximize not only stiffness, but also maintain shape. If these plys were "laminated" randomly, or all in one orientation, then the woven fibres would fight and pull against each other and force the finished component out of shape. In some instances, "UD" (unidirectional) carbon fibre is used to deliberately bring stiffness in one plane only (or reduce stiffness in another plane). The end product is then cured in an autoclave ready for use.

Composite Laminating Principle using Carbon Fibre or CFRP (Carbon Fibre Reinforced Polymer)

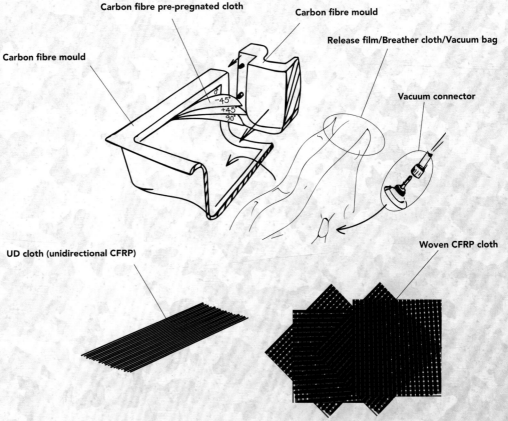

Carbon fibre pre-pregnated cloth

Carbon fibre mould

Carbon fibre mould

Release film/Breather cloth/Vacuum bag

Vacuum connector

−45°
+45°
90°

UD cloth (unidirectional CFRP)

Woven CFRP cloth

HONEYCOMB CARBON FIBRE

Solid carbon fibre is a monolith of material, and indeed very strong, but as Gordon Murray states, the real multiple benefits come with so-called "honeycomb" carbon fibre. Two sheets of carbon fibre cloth sandwiching Nomex or aluminium honeycomb bring enormous stiffness and lightness. Also, other core materials and inserts can be incorporated so that mounting an engine off the back of a chassis is comparatively straightforward.

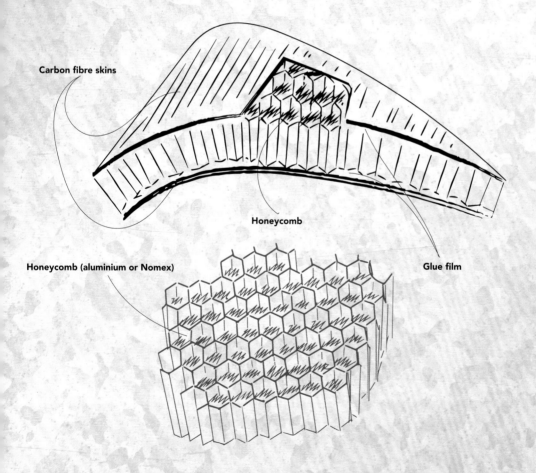

Carbon fibre skins

Honeycomb

Honeycomb (aluminium or Nomex)

Glue film

By the end of the nineties, the supercar world had recovered
somewhat from the devastating recession and over the next
few years the marketplace became very crowded. The choice
on offer was huge and new manufacturers vied for the big
cheques that eager supercar collectors would gleefully hand
over. However, so advanced was the science of these machines
that the development costs were now terrifyingly high. Indeed,
so prohibitively extortionate was the R&D that new supercar
marques tended to come and go in a mess of financial problems.

Nonetheless, as the new millennium approached, a
remarkable development in the world of supercars shook up
the establishment, against all the odds. One new Italian brand,
Pagani, came along with a car that was so brilliant and incredibly
well thought out that it took the supercar world by storm. If the
F1 was the moment when supercar science made a wholesale
shift to the extensive use of composites, Pagani escalated that
science into both a performance essential and also an art form.

Pagani had never previously made a road car when they
launched their stunning Zonda in 1999. The Argentinian founder
of the new marque, Horacio Pagani, is the son of a baker. He was
drawing car designs even as a young child and constructed his
first buggy car aged just 16, followed by a single-seater F3 car
in his early 20s. Moving to Italy with a letter of recommendation
from the Argentinian racing legend Juan Manuel Fangio, Pagani
got a job at Lamborghini, where he immersed himself in the
evolving science of carbon fibre. By 1991 he had set up his own
carbon fibre manufacturing centre in Modena, making parts for

car brands, including the Ferrari F1 team. He began drawing ideas for an entirely new supercar and created that vision himself using old-school draughtsmen's pencils and drawing boards. When his company (consisting of only 70 people) revealed the Pagani Zonda at the Geneva Motor Show in 1999, complete with a 7.3-litre, 450bhp V12 AMG engine, the car world took a sharp intake of breath. The car was so striking, so beautiful and so scintillatingly fast that everyone had to take notice. Apart from an aluminium subframe, the rest of the car was constructed from carbon fibre. Remarkably, the interior finish was also almost entirely carbon fibre and was finished to a high level of craftsmanship more akin to a jeweller's shop window than a car cockpit. Curved, three-dimensional objects such as the steering wheel were created from carbon fibre, yet every single one of the thousands of strands used would meet in perfect unison, a hand-crafted job done entirely by the human eye. This level of obsessive precision extended to every super-tight weld, every nut and bolt on the Zonda and elevated the science of the supercar to a new level of precision-meets-art.

Horacio's brilliance is the result of a very personal vision that sees art and science perfectly merged together – one of his biggest inspirations is Leonardo da Vinci. He is best placed to explain how he approaches the use of carbon fibre in his creations: "In the early eighties, when I worked at Lamborghini, I had the opportunity to become familiar with these materials and understand their potential: the possibility to create shapes, even highly plastic ones, with a material featuring very remarkable structural characteristics and which weighed very little. I was able to compare carbon fibre and Kevlar with aluminium and steel, and I saw how much potential there was in these newer materials. Of course, there were huge difficulties in how you used the materials for the bodywork, but let's say I immediately

realized that, although the route was difficult, these materials would become a key to success in the future.

"To begin with, many designers who were used to working with traditional materials found the transition somewhat traumatic, since their approach was to replace something made with steel with something made with composites. For me, as a designer, it was perhaps more straightforward, because I wasn't bound by one technology or another. For me, each technology was equally valuable, and this gave me the freedom to create without constraints. So I tried to learn as much as I could about the potentialities and limitations of the material, and as I gradually became acquainted with it and its technology, an incredible scenario opened up, allowing me to create the shapes I dreamed of."

Horacio Pagani's own fascination with art history and philosophy means he is well placed to comment on how supercars might strike a refined balance between art and science: "Well, let's say this is a school of thought which in some way Leonardo da Vinci brought to us over 500 years ago, one I humbly took on board, and which we try to pursue in the same spirit at Pagani: I believe the balance is 50–50, and let's say that for us it's normal that when we work, we imagine something which must have a very specific function but must also look good, must express something, some message. So, whether we design a suspension, a wing-mirror or an inner part, one that is visible or one that isn't, we always use the same approach. It makes no difference to my team and me, we always start from

the notion of creating the best there is, from a technical and technology point of view, but also in terms of aesthetics."

Perhaps Horacio Pagani's greatest achievement was to launch a successful new supercar company in a modern world that was dominated by huge international corporations. However, it is his unrivalled use of carbon fibre and composite materials that is arguably his greatest contribution to the science of the breed. Yes, high performance and luxury cars had been using exotic materials to create a superlative finish for decades. However, it was the fact that the Zonda could make composites look so stunning while also contributing massively to the car's light weight, performance and structural integrity that was Pagani's *coup de grâce*.

As both the McLaren F1 and Pagani Zonda exhibit, advancing the supercar genre represents a core balance between the search for more power versus the need for less weight. Gordon Murray has a fundamental problem with the breed of heavier modern supercars: "That isn't pure. To me, to call anything over 1,500kg [3,300lb] a sports car is sacrilege. It is fair to say that many of these heavier supercars don't feel like two tonnes. However, when you get on the small roads around the Targa Florio and you try and chuck it into a hairpin, it just wants to keep going in the direction of the mass and the transient handling is very lumbering and elephant-like. There is a massive difference between power-to-weight and weight-to-power ratio and that is a fundamental thing that some people, including car designers, just don't understand."

OPPOSITE AND ABOVE_In the Pagani Huayra Roadster the use of carbon-fibre was taken to arguably unrivalled heights.

POWER-TO-WEIGHT RATIO/ WEIGHT-TO-POWER RATIO

Power-to-weight ratio is the power in the engine in relation to the weight of the car. Simply put, this relationship affects acceleration and handling. Supercar designers seek to minimize weight and maximize power to increase the driving experience and thrill. It is expressed as horsepower divided by the weight of the car (empty) in pounds. Therefore, for example, the exoskeletal, super-lightweight track car the Ariel Atom 500 V8, has a power-to-weight ratio of 900bhp per tonne; the first Porsche 356 had a power-to-weight ratio of 47bhp/tonne; the first Porsche 911 had a power-to-weight ratio of 120bhp/tonne; a modern-day 911-991 Turbo S has a power-to-weight ratio of 346bhp per tonne.

Weight-to-power ratio is an expression of how much needs to be done by each horsepower. A 1,001 horsepower engine in a car that weighs 5,500lb gives a ratio of 5.49lb per horsepower (weight divided by power). The lower this number, the more the rate of acceleration increases.

McLaren 720S

LEFT_An expanded cut-away of the Bugatti Veyron's skeleton.

"It doesn't matter what car you are designing," continues Murray, "whether it's a one-litre city car or a hypercar, *weight* is the one thing that counts against a designer every single time the car is not standing still. If you are accelerating, weight counts against you. If you are braking, weight counts against you. And I'm not just talking about braking distance, I'm talking about weight transfer, the way the car moves and pitches. Transient handling, cornering, change of direction, every time the car is moving, weight is your worst enemy."

This is why Murray and Pagani both maximized the use of composite materials in their cars and why, in the aftermath of the F1 and Zonda, every other supercar manufacturer of any note spent millions of pounds evolving and honing their use of these lightweight, super-strong materials. Even after the F1, some supercars still used steel for the subframes or various crash structures, but the relentless push for a comprehensive composite structure was unrelenting. Ferrari was the first competitor to follow the path of the McLaren F1 when they used a full carbon chassis on their F50 in 1995. Then in 2002, Ferrari released the Enzo, which used carbon fibre for all of its bodywork and structures. Since that point, the practice has become almost universal and to make a modern supercar that can match the performance of its contemporaries without comprehensively using the science of composite materials is now unthinkable.

"Every time the car is moving, weight is your worst enemy."

Gordon Murray

CHAPTER 9

SHIFTING GEAR

By the year 2000, supercar science had evolved to such a high level that any car not capable of 200mph+ in relative comfort, with ample technology on-board to help with traction control, power assistance and electronic systems management, all complete with stunning looks, was considered a failure.

OPPOSITE_The Bugatti Veyron obliterated all supercar competition on its 2005 launch and has come to be seen as the benchmark for all modern hypercars.

To deliver a game-changing new piece of supercar science was becoming increasingly difficult for manufacturers. Luckily, the thirst for supercars seemed to be growing every year, with more new models being released and special edition cars being snapped up by frantic collectors, all fuelled by an explosion of magazines, websites, auctions and events featuring supercars in minute detail.

Despite this enthusiasm, the environment for producing a watershed supercar was not particularly inviting. By the end of the 20th century, car companies had been consolidating at pace, as larger corporations bought up smaller marques en masse. VW had added Lamborghini and Bentley among others to their stable of Seat, Audi and Skoda; BMW bought Rover and Rolls-Royce; Ford acquired Land Rover, Jaguar and Aston Martin. With these mergers and buy-outs, the pressure was truly on to justify a car's scientific development costs, because board meetings

RIGHT_Bugatti Veyron Super Sport, 2010

were home to more accountants than car designers or engineers. The so-called supercar halo effect had to have more than just some nebulous street credibility; financial numbers mattered more than ever. Economies of scale were prioritized within a group budget to please shareholders, so the days of extravagant supercar development for the sake of it seemed long gone. Also, the pressure on emissions was reaching fever pitch and compounded the social disdain towards supercars that in some quarters had never dissipated. The question was, in this new era of superlative machines and massive public interest, who could produce a genuinely game-changing supercar and what on earth would that car need to be capable of?

"Who could produce a genuinely game-changing supercar?"

The answer came with the launch of the Bugatti Veyron in 2005, by a French manufacturer that was now part of the world's biggest car group, VAG (better known as Volkswagen). The Veyron was the brainchild of Dr Ferdinand Piëch, the grandson of Ferdinand Porsche. His dream had been to produce a car with the same 1,000bhp that he had seen during his time working on the legendary Porsche 917 world championship-winning race car.

ABOVE_The legendary Porsche 917 used 1,000bhp and piqued the interest of engineering genius Dr Ferdinand Piëch, who was determined to use the same amount of power in a road car one day.

OPPOSITE_Various Bugatti prototypes, displaying the lengthy gestation period for the astonishing Veyron.

He had also been a part of the Quattro project, so mighty grip was paramount. Yet he also wanted a car that was so precisely engineered it could be "driven to the opera". A top speed in excess of 250mph and a 0–60mph time of less than 3 seconds were the final parts of the brief – scribbled, according to Bugatti legend, on a small napkin.

The brief was short but the task – and the science – would prove somewhat less simplistic. The Veyron project battled through a long gestation hampered by protracted technical issues, simply due to the colossal goals that had been set. At one point the test cars were simply not working as planned and some doubters wondered if the car would ever go into series production. However, with a team led by Dr Franz-Joseph Paefgen and Dr Wolfgang Schreiber, the car was finally revealed in 2005, some £1.5 billion ($2.1 billion) worth of research and development later.

The Veyron was worth the wait. The standard Veyron had 1,001bhp, generated by a 16-cylinder 8-litre engine. The

BELOW AND OPPOSITE_ Such was the power and performance of the Veyron Super Sport that conventional supercar science had to be re-invented.

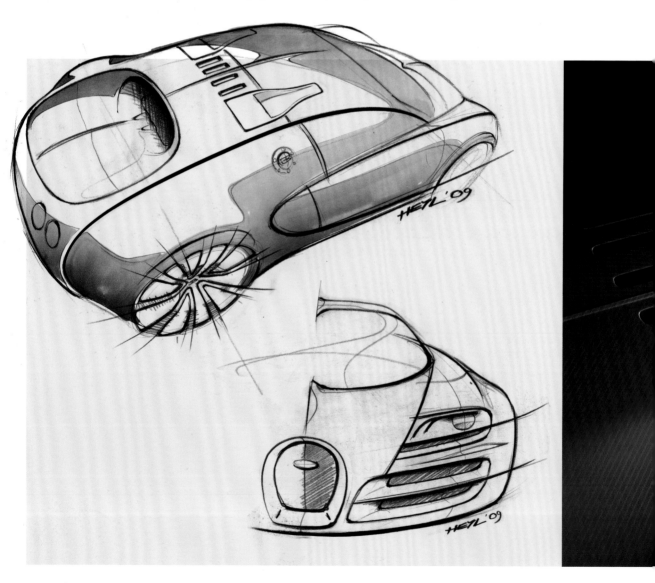

"The Veyron was worth the wait."

Veyron's statistics were so far superior to the competition that it redefined the genre overnight. In terms of performance figures, the McLaren F1 had finally been usurped. The 0–60mph time of 2.5 seconds was a world best; the top speed would eventually hit 267.856mph in 1,200bhp Super Sport trim (with 1,200bhp); the standard Veyron hit 200kmh (125mph) in just 7.3 seconds and 300kmh (186mph) in a staggering 16.7 seconds (the later more powerful Super Sport version hit these speed landmarks even more rapidly, in 6.7 seconds and 14.6 seconds respectively). Perhaps even more remarkably, the car could indeed be driven easily and gently around town.

The science behind the Veyron was staggering. On one hand, much of the tech had been used in cars before – a carbon fibre monocoque; lightweight materials wherever possible; turbos (the Veyron had four); custom lightweight wheels; intelligent all-wheel drive; intricate electronic systems management; even 16-cylinder engines (the Cizeta-Moroder in 1988). However, it was the *degree* to which the Veyron designers utilized these technologies that set it so far apart from every other supercar on the planet.

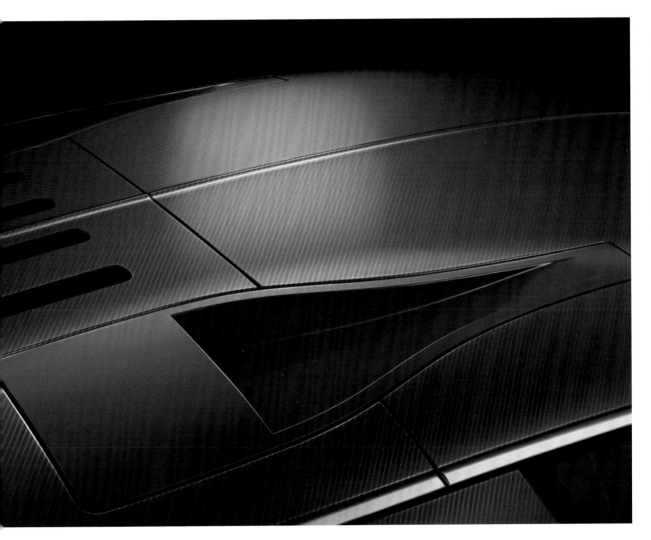

THE DIFFERENCE BETWEEN HORSE-POWER AND TORQUE

Torque and power are expressions of force and work done. Horsepower is an expression of the rate at which energy is transferred to move an object over a given distance in a given time. The concept was first developed in the 18th century by the Scottish inventor, mechanical engineer and chemist, James Watt (whose surname describes another expression of power – the watt). Supercars will have a theoretical maximum power figure reported in horsepower; brake horsepower reports the actual measured power after friction losses. Other terms are sometimes used to express power, but essentially all are based on the principle of the quantities of force, weight, distance and time (i.e. "work done").

Torque is a moment of rotational force that the pistons place on the crankshaft (the explosions in the cylinders give the pistons downwards movement and these rotate the crankshaft via connecting rods). Rotational force is defined by the mass and acceleration. A simple analogy: imagine sitting in your car and opening the door by pushing on it close to the hinges – difficult? Whereas if you push the door at the point furthest away from hinges, it's a lot easier. This is torque, a rotational force.

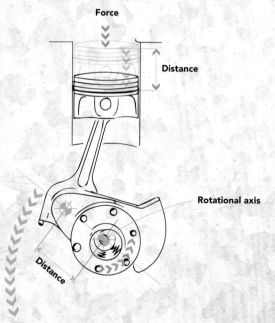

Force

Distance

Rotational axis

Distance

Torque

>>>> Horsepower is an expression of the rate at which work is done
Horsepower = Force x Distance x Engine Revolutions per minute (RPM)

>>>> Torque is an expression of Twisting Force
Torque = Force x Distance

Reviewing this famous supercar over a decade after its initial launch, there is one particular element that perfectly summarizes the Veyron's triumph and, arguably, was the single piece of science that made the car's astonishing performance possible: the so-called DSG gearbox. With Dr Schreiber at the helm as chief engineer, the Bugatti team were faced with an engine so powerful, creating forces so gargantuan, that conventional gearboxes were being, quite simply, shredded. The forces being blasted through the transmission were unprecedented in a road car. However, it was not good enough to simply use a sturdy racing gearbox as this would have been impractical for the genteel road use that had been stated in the initial brief by Dr Piëch. Existing race car paddle shifts were too abrupt and harsh for use in this super-luxury high-performance road car. Similarly, existing road car paddle-shift systems were considered far too slow and clumsy, potentially stunting the stellar acceleration with slow shifts.

Gearbox science had, to some degree, lagged behind other supercar tech for a while. Ironically, the way gears affected performance was one of the very first elements of science that engineers looked at improving when the car was first invented. When Karl Benz was developing the first ever car, the Motorwagen, initial test runs had proved very sluggish and unreliable so he inserted a second gear to improve speed. The contemporary duo of Panhard-Levassor were also pioneers – a twin-cylinder model of 1899 had two-gear levers, highly unusual at the time. Many view them as the forefathers of the modern gearbox, although one reporter in the late 1800s said their revolutionary ideas were "more hocus-pocus from charlatans trying to cash in on the public's fascination with the new motor car". Louis Renault evolved the system further and created a shaft to the rear axle (rather than the earlier chains). Variations

OPPOSITE_The
Veyron Supersport
World Record
Edition – only
nine examples
of this unicorn
of supercars were
built, following
Bugatti achieving
a world record
speed of over
267mph.

on these key early ideas have largely been adopted by the majority of the world's manufacturers in some form in the century or more since.

The advent of the automatic gearbox changed the game, with the very first primitive example arriving in 1904, designed by the Sturtevant brothers of Boston. In the late thirties Oldsmobile introduced a more advanced four-speed automatic transmission, along with other contemporary innovations by General Motors, Buick, Chrysler and many others, and ever since America has been the principle champion of the automatic gearbox. By the fifties, so many other countries were adopting this easier-to-use technology that many were predicting the extinction of the manual gearbox altogether.

However, in the late eighties, the Ferrari Formula 1 team began developing a gearshift system that used paddles on the steering wheel instead of a conventional gear stick. The idea was to create easier and faster gear changes to advance their racing cars (and also allow narrower cockpits). The testing was done by the actual F1 drivers of the time, including the legendary 1992 F1 World Champion and 1993 Indycar World Champion, Nigel Mansell: "It was 1989 and I was testing this new system directly on track. They had loads of issues with it, the technology wasn't initially working, so it was a very intense time. We had so many failures. When it did start to work, it was very good, but at the same time it was very slow at changing gear.

A big issue were the solenoids in the gearbox, which kept failing. There were also issues with the wheel sensors and if they failed too, it could get very challenging.

"Gearbox science had, to some degree, lagged behind other supercar tech for a while."

"I had come from a time when race cars had stick shifts so I had to get accustomed to this new system at very high speeds. The stick shift system of old was a hell of a lot more reliable and, to be honest, I enjoyed pulling the gear stick and changing the gear in that way when I wanted, with an immediate response. With the semi-automatic, you'd use the paddles, then have to wait for it to go in gear. Sometimes it didn't even go into gear. It was a massive task to get the whole thing working properly – back then we didn't have complex simulators, so we'd be out on track on a weekly basis.

"During pre-season testing both my car and Gerhard Berger's car never completed more than about five laps. However, the most amazing thing was that when we went to the season opening race in Rio, this car worked, didn't break down and I won the race! It took another eight races for it not to fail again, but that first win was pretty special. I went through similar anguish in 1991 when I joined Williams because of a semi-automatic gearbox which was so unreliable. However, you have to remember that these engineers were pathfinding and pioneering, so that was to be expected."

The first road car to adopt this F1 technology was the subsequent Ferrari F355. Just like the F1 car of Mansell, there were early problems: shifting was often sluggish and drivers

reported a lack of "feel" when making gear changes. Worse still, some cars made shifts that were juddery and at times violent, as the struggling technology crunched its way through the cogs. Given the imperfections, manufacturers at that time almost always offered a conventional manual gear shift as an alternative to the paddle system because the technology was not considered consistent enough, plus many customers were not convinced it was superior. The science on offer was initially flawed, but the system's gradual adoption by race teams around the world led to the problems eventually being ironed out.

By the time the Veyron was being developed at the end of the nineties, the power produced by its eight-litre, W16 configuration engine was so vast that, as noted, conventional gearboxes were quite literally being obliterated. Enter the DSG gearbox (the letters stand for *Direkt-Schalt-Getriebe*, or Direct Shift Gearbox). The DSG is a double clutch system that is electronically controlled to produce hugely fast gear changes and, due to its construction, is extremely robust. There is no clutch pedal, even when using the manual paddle-shift levers on the steering wheel, creating both fully automatic mode as well as semi-manual with the paddles – hence the ability to race the Bugatti at full tilt with manual paddle shifts or alternatively cruise around town at low speed in full automatic mode. The DSG is capable of superfast gear changes – under 150 milliseconds, which is quicker than the blink of an eye. This science eliminated the crunching, jarring gear shifts of previous systems and was both quicker and far more robust. One downside was that DSG boxes could be rather heavy, but the pay-off in terms of performance was massive. Crucially, the Veyron's DSG box was also bulletproof.

BELOW_The Bugatti Vision Gran Turismo one-off hypercar.

THE TWIN CLUTCH DSG GEARBOX

To simplify the complex science enormously, the DSG box is effectively two separate gearboxes and clutches in one coherent unit. The secret is the use of two input shafts instead of one (with a clutch managing each shaft), meaning that multiple gears could be engaged at the same time – essentially the gears before and after the one the car is in – in preparation for the next shift. Two "wet" clutches (immersed in oil for cooling) are used to activate one of two concentric input shafts (concentric meaning that one fits over the other). One shaft carries odd gears (first, third, fifth and seventh), the other the even gears (second, fourth and sixth). When a gear is selected – say third gear –

then the gears on either side of it (on the other shaft) are simultaneously pre-selected, so that when either fourth or second gear are called for, these next gears are ready to be used – the only remaining mechanical action is clutch one handing over to clutch two. This does away with time delay, as the gear movement has already been dealt with – because while driving along in third gear, second and fourth are active, but not yet transmitting power to the output shaft. Further, there is a torque transfer overlap between clutch one and clutch two, making for a very smooth handover from odd to even gears and vice versa. This results in almost instantaneous and sublimely smooth gear changes.

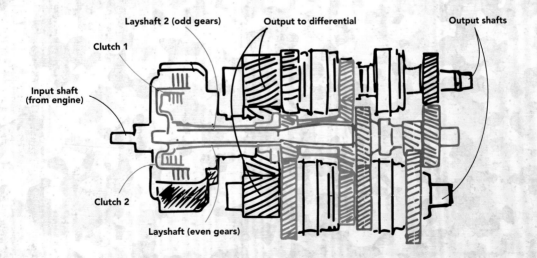

Layshaft 2 (odd gears)　　Output to differential　　Output shafts

Clutch 1

Input shaft
(from engine)

Clutch 2

Layshaft (even gears)

Concentric layshafts and clutch assemblies, one sitting inside the other, is where the magic happens.
Odd and even gears are selected and pre-selected, with only the clutch engagement needed between
each gear change (as next gear is already selected).

STEERING WHEEL
AND GEAR SHIFT
TRIMMED WITH
VELVET

Michael Kodra is head of the transmission development at Bugatti, having previously worked at VW Motorsport, including on WRC Rally cars. "For sure, the DSG is a significant advantage in the whole design of both the Veyron and the Chiron. If you sit in an ordinary supercar with a standard gearbox, then you have a clutch pedal on the left, but due to the size of the clutch needed to handle the torque of these supercars, the feeling and sensitivity of that pedal is usually not very easy, even for experienced drivers. When making the Veyron, it was clear that the DSG would enable the car to be driven by anyone and that was absolutely crucial in its development. The team could have designed a standard single clutch gearbox to cope with the huge torque, that was not even necessarily overly difficult. However, the driver would still have needed to find the correct biting point and so on. Whereas the DSG uses a similar sized clutch (only slightly bigger), but the system controls that using hydraulics, not the driver's foot on a pedal. We also used standard high precision steel because no other material could reliably cope with the massive forces being generated."

This was a fine balance, as too much assistance might have rendered the Bugatti feeling too robotic. To avoid this and ensure the driver always felt in control of a

"When making the Veyron, it was clear that the DSG would enable the car to be driven by anyone."

Michael Kodra, head of transmission development, Bugatti

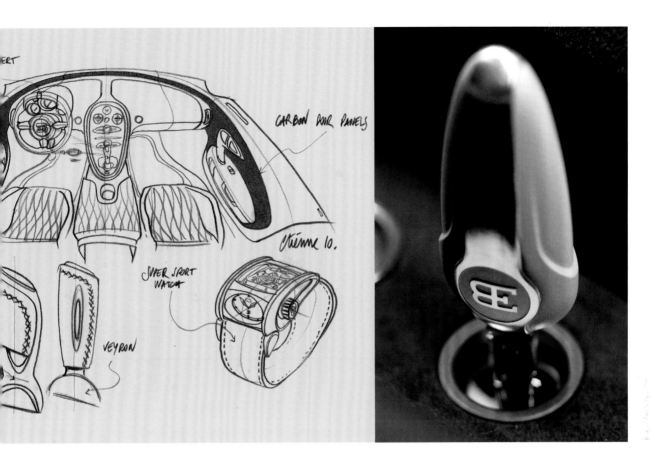

The following labels appear in the sketch: CARBON DOOR PANELS, etienne 10., SUPER SPORT WATCH, VEYRON

rewarding experience, Bugatti fitted various settings ranging from manual (using the paddles on the steering wheel as if it were a fully manual car) through to fully automatic. Yet they were able to do this with a DSG gearbox – it was not unusually large or heavy. "The gearbox itself," continues Michael Kodra, "and the gear-set is similar in size to standard single clutch systems – there are still shift forks and synchronizers and the gears, much of the internal workings are very similar and so the basic layout of the gear-set itself is like a manual transmission."

The ingenuity of the DSG gearbox is that it was able to cope with the gigantic torque of this record-breaking car without being too large or too heavy, while at the same time being very easy for a wide range of people to enjoy driving. By overcoming these engineering challenges, it could be argued that the DSG is possibly the biggest single development in the history of gearboxes since the very first single clutch systems were introduced over 100 years ago.

If the DSG gearbox made the Veyron's astonishing performance figures possible, it is also worth noting how supercars of this calibre are able to transfer the enormous power generated on to the road surface. Supercars certainly did not invent the tyre, nor did they play much of a role in the early evolution of the science behind tyres, that story is for another book. However, as the supercars on sale have become faster and

ABOVE AND OPPOSITE_Without the super-strong DSG gearbox, the Veyron would have struggled to put its apocalyptical power down onto the road.

TYRES

Tyres need to squash, bend, go through extreme heat cycles, support load and give grip and traction in all weather and across a huge range of conditions. They must achieve all of these characteristics by exhibiting predictable, repeatable behaviour and return minimal wear.

The primary materials used in tyre construction are steel, Kevlar, nylon, silica and carbon black, along with speciality fibres. Typically there's an extraordinary number of raw materials used, frequently more than 150. Supercar tyres have features not afforded to "normal" tyres. For example, the rubber compounds in the tread of a supercar tyre will differ from outer edge to inner, because outer tread is under greater demand during cornering. Tyre contact patch is all-important for a supercar. The bigger the contact patch, the more the driver will extract from the engine power, but to maximize this potential, tyre pressures should be run as low as the tyre walls will enable. Tyres need compliance and "squash" to operate effectively. Ironically, low-profile tyres are not optimal in terms of high performance.

Supercar tyres are tested to destruction using aircraft tyre test dynamometers and rigs. Track testing of supercar tyres is exhaustive and the process is also heavily reliant on vast computing power, as well as chemical and mechanical engineering, along with the ever-improving understanding of the influence of the laws of science.

Cross-section through Tyre

O **Tyres stationary**
Greater contact with road surface

O **Tyres running at 250mph**
Tyre walls have narrowed, tyre radius increased

more powerful with every passing year, the demands made on the humble four corners of rubber have increased exponentially.

A standard road car tyre has no more rubber touching the road than on the soles of your feet. Even allowing for much wider and bigger supercar tyres, the contact patch is still relatively small – in the region of 85cm^2 (13in^2) on the front and 130cm^2 (20 in^2) on the rear for a Bugatti Veyron.

Acclaimed racing driver Andy Wallace was immersed in these tyre challenges when he broke the world production car speed record in the McLaren F1: "I think we probably spent about four hours making adjustments and measuring the edges, making sure everything was good on the tyre front. Fortunately Michelin's development facility was incredible. The forces they can put through a tyre to be sure that it can reach its designed parameters are unbelievable. If Michelin say to you, 'This tyre is good for this speed', then you know that is absolutely the way it is."

The F1's designer, Gordon Murray, explains why the tyre is a crucial part of any supercar science and how there are knock-on effects all the way through the car's set-up: "Tyre design, such as the side wall of a tyre, is so much part of driver feel. If you take stuff like suspension geometry, steering geometry, steering feel, steering ratio, static mass distribution, centre of gravity, all that stuff, and put that all in a bucket, then in terms of feel back to the driver, I'd say more than half is just in the side wall of the tyres. Tyre designers hate heavy cars … these modern supercars are so draggy, most of them don't do over 210mph because they are just big bricks – they just drag. They just go after downforce, but tyre designers hate it when you have a heavy car and a fast car. For a start you have got massively compromised side walls and that affects everything in the car, which you then have to overcome with electronics, power steering, ESP and traction control, all that sort of stuff, to overcome that lack of feel. Well, you don't overcome it, but you disguise it."

Back in north-eastern France, for the Bugatti Veyron to be safe to sell for series production, the team needed a tyre that was capable of driving at 30mph to the opera in comfort but also able to hit the standard car's top speed of 253mph *at any given moment*. That meant a tyre capable of speeds around 70+mph faster than a Formula 1 car – quite a task. That is why the bespoke Michelin PAX tyres were unique to the Veyron, could only be removed at a special facility in France and cost £18,000 ($25,000) a set (they could also be driven when flat).

Even though Michelin produced a road tyre that was significantly superior to anything that had preceded it in the supercar world, at 253mph the forces are so enormous that the rubber will only last for around 15 minutes. That figure is actually irrelevant because at that speed the Veyron's petrol tank will be empty in 12 minutes.

"The Michelin PAX tyres were unique to the Veyron and cost £18,000 ($25,000) a set."

CHAPTER 10
THE ART OF AERO

Many of the scientific advances in this book have appeared rather suddenly – at least in terms of the car they were launched on.

So it can be stated that the first independent suspension on a supercar was the Gullwing; the first mid-engined supercar was the Miura; the first carbon fibre monocoque was on the McLaren F1. However, one of the single most important scientific elements of any supercar is also a field of science that has seen mostly incremental improvements over the years: aerodynamics.

Scientists have known for many decades how the flow of air affects vehicles, but as an exact science it has taken a long gestation for aerodynamics to be applied to high-performance automobiles with real *precision*. It has already been detailed how very embryonic aero influenced the first high-performance cars and land speed records at the start of the 20th century – albeit with sometimes questionable results. Even after World War II, the automotive industry's understanding of aerodynamics was modest, with primitive wind tunnels and no computer aids to assist in the pursuit of aero efficiency. As has been noted, in the immediate post-war years aero became unfashionable as Americans moved towards large, luxurious cars while Europeans moved towards economical mini vehicles at a time of rationing and austerity.

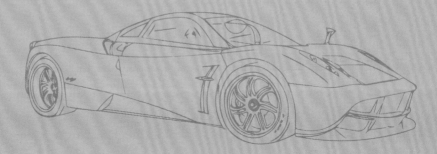

RIGHT_ Pagani Huayra, 2012

As mentioned earlier, the wedge supercars of the seventies
might have caught the eye, but also a lot of drag. Nonetheless,
aerodynamics eventually returned to the drawing board of
the world's automotive engineers in the seventies, partly as a
response to the fuel crisis of that decade – and everyday cars
were at the forefront as much, if not more, than supercars. One
emerging contemporary vogue within mass-market cars was for
so-called "detail optimization", which essentially meant focus
was placed on small, individual elements such as the camber of
panels, side mirrors, fairings, the edges of pillars and so forth.
Although cars' drag coefficients continued to fall in mass-market
vehicles and high-performance breeds, at this point there was no
comprehensive science behind the entire car's aerodynamics.

It quickly became apparent that detail
optimization had limited scope and when
drag coefficients ceased to fall, engineers
in the car industry actually looked back at
the history books and found the thirties
was a decade that might lead the way
forward. This period prioritized shape
optimization, whereby the overall shape was the core element
of drag reduction, which in turn was then honed and perfected
for even lower drag results. Designers realized that using this
approach in conjunction with shape optimization was crucial to
provide a more coherent and effective pursuit of aerodynamic
efficiency. However, supercars still cannot claim credit – much of
this science was on mass-market cars. The first car to fully utilize
shape optimization was the Audi 100 in 1983, which had an
unheard of drag coefficient of just .30 (less than a McLaren F1).

**"The first car to fully utilize
shape optimization was
the Audi 100 in 1983."**

AERODYNAMICS

Drag is a resisting force created by a car's form as it passes along the road. Too much drag can affect performance. Drag is a quantifiable value that can be measured (for example in kilogrammes). Imagine a cube travelling through water – it's "draggy", right? Now imagine a teardrop shape with the same area doing the same. The teardrop shape is far less draggy. One of the factors used in calculating a car's aerodynamic efficiency is drag coefficient – a car with a high drag coefficient has generated a lot of drag. A car with a low drag coefficient has proven to be slippery through the air and met with less resistance. Calculating and manipulating these forces is considered one of the finer arts of supercar and race car engineering

(see page 162). Computational Fluid Dynamics (CFD) along with testing in wind tunnels are the chief methods of finessing a supercar's drag coefficent and aerodynamics.

Supercars have to strike a greater balance than race cars in this area. Formula 1 cars are horrifically inefficient in terms of drag because they generate such massive amounts of downforce and have open wheels (no bodywork or fairings over wheels and tyres). Although supercars also need to generate downforce so they can corner and perform at high speed, they are not subject to motorsport regulation, which means they can use active aero to change the aero surfaces (the wings) when maximum downforce is not required. Downforce always comes with a drag penalty.

Wind Tunnel
_001 Turning vanes
_002 Chiller
_003 Contraction (4:1)
_004 Boundary layer removal duct
_005 Rolling road
_006 Supercar (full size)
_007 Breather slots
_008 Fan motor (electric)
_009 Fan blades
_010 Settling chamber screens

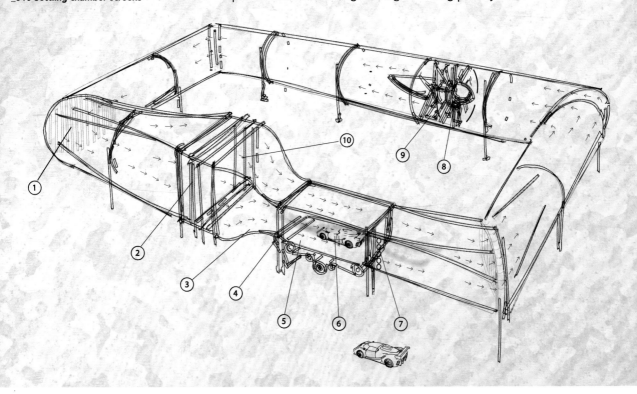

None of these advances in the modern era would have been possible without the aid of a tool culled from the aviation industry – the wind tunnel. As noted earlier, although wind tunnels had been in use for some years, it was not until after World War II that motor racing teams started maximizing their benefits. Inevitably this passed down to supercar tech, although the high costs of wind tunnels meant that only the very wealthiest marques could afford their own facility.

Tony Hatter, the man behind the Porsche 911-993 Turbo, has some revealing insights into the use of wind tunnels and also the internal dynamics of producing a supercar: "The Weissach Development Centre was unique at that time, combining not just the design studio and workshops but also the test track, prototype facilities and the wind tunnel. This meant we could work hand in hand with the aero guys. As the design department, we came up with the shapes and forms in the first place, then it was taken to bits by aero or production, we always provided the goods and then all these different departments with their own goals got involved. It is always a constant discussion. We'd have a meeting every month and often the aero goals weren't met because studio wanted a certain design feature, then either we'd win because it looked better or the aero guys would win because they'd gain a consumption improvement or similar. A lot of things go on that influence what you are doing on a car. The ergonomic people come up with numbers and figures, but all we can do as stylists is say, 'Yes, but this looks better!' It can be a very difficult corner to defend. A juggling act.

"For the 993, we started off with a basic 911 and had to modify it to be able to perform as a Turbo, so faster, bigger wheels, a

ABOVE_The Porsche 993 RS, a classic 911.

OPPOSITE_The American Le Mans series-winning Acura was created entirely using Computational Fluid Dynamics, courtesy of Wirth Research.

spoiler and so on and at the time we thought, 'Wow, 400bhp, this is the limit!' Our main challenge was the aerodynamics as well as thermodynamics, getting enough air into the engine and getting enough downforce to make it stable at speed. In many ways it was an aero project."

With modern-day supercars routinely exceeding 200mph, the need to be aerodynamically refined has never been more apparent or necessary. The value of the wind tunnel has been somewhat diluted, too. It is arguable that it was not until its full immersion in the world of computer-aided development that the supercar truly began to master aerodynamics. With the advent of computer technology, it was soon possible to produce air flow calculation programmes that were minutely accurate. Once more, motorsport led the way, with pioneers such as Nick Wirth creating the first race cars entirely using Computational Fluid Dynamics (CFD), a case in point being the American Le Mans series-winning Acura. Wirth then took this tech into Formula 1 with Virgin Racing and has since established his eponymous company as a world leader in the field.

> "The need to be aerodynamically refined has never been more apparent or necessary."

COMPUTATIONAL FLUID DYNAMICS

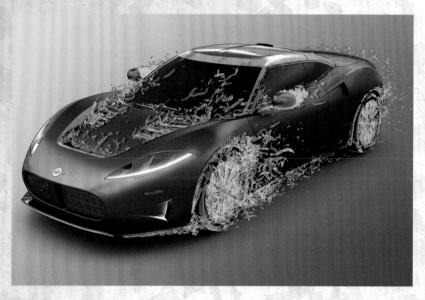

Above_Wirth Research carried out CFD work on the Spyker Preliator using a supercomputer with 4,000 processing cores. In 2007 a racing car CFD run using the two cores available, running 15,000,000 cells (the increments into which CFD models are divided), took 300 hours. Using Wirth's latest technology, 400,000,000 cells on 300 cores multiplied by 11 different models can be run simultaneously.

CFD is powered by massive computer power calculating sophisticated mathematical algorithms and numerical analyses to predict the behaviour of airflow over and around an object (in this case, a supercar). CFD gives a greater understanding of flow and how everything works together, as Wirth Research's Engineering Manager Rob Rowsell explains: "CFD offers flexibility in workflow. With wind tunnel testing there is a lag as ideas are schemed and developed then physical models made before being placed in the wind tunnel. The benefits of CFD extend beyond just time saving. The physics can be better understood and altered; for

example, conditions inside brake duct cooling and radiator ducts, even air-conditioning can be run through CFD. The software also gives a much better feel for why a change had a particular effect.

"Images, plots and movies are generated to display the findings and help give direction for the next set of test criteria and conditions. Experience and knowledge is then needed to carry out 'post-processing' as part of interpreting the digital results. Further, cross-referencing results using CFD, wind tunnel and ultimately track tests for correlation and verification is still an important part of thorough testing best practice." These physical methods such as track tests and wind tunnels will yield results such as measured forces and targeted pressure readings, but these are effectively snapshots, whereas CFD can predict based on transient conditions, such as the supercar pitching, rolling, accelerating, braking, yawing, sliding, making demands of the brakes, cooling, steering and its aerodynamic package. It is important however that the CFD results are correlated to experimental and real-world data. Wind Tunnel Models and/or Track/Road Test data and results are fed back into the CFD algorithms and "recipes", thereby enabling these super-sized calculations to become ever more accurate and effective.

There are obvious supercar highlights in the science of aerodynamics, the McLaren F1 being an obvious example: "It was going to be quick," explains its designer Gordon Murray, "because it was very small for a supercar. It had a drag efficient of .32 … I didn't want any obvious wings and flaps and spoilers. I wanted the classic shape that would still look good in 50 years' time. I got slated a bit by a few journalists when it came out that it wasn't dramatic enough, but I didn't care because I knew what I was doing with the shape."

There are many other examples of brilliant aerodynamics on modern supercars that could be used to best exemplify the art – the Ferrari Enzo spent months and months ensconced in a wind tunnel and actively utilizes so-called adaptive aero to create downforce and grip, a tech that had previously been banned from Formula 1. Other notable supercars such as the Pagani Zonda, the Bugatti Veyron and Chiron all demand aerodynamic plaudits on a list that is long and impressive.

Somewhat spoiled for choice on which supercar to examine to explain aerodynamics, for the purposes of this chapter the stunning new Red Bull Aston Martin Valkyrie will be used, principally because its chief designer, Adrian Newey, is a man widely regarded as the guru of all things aero.

Codenamed the AM-RB 001, the Valkyrie takes its appellation from Norse mythology, continuing the Aston tradition of names beginning with V. Newey has worked with Aston Martin designer Marek Reichman to produce a useable road car that is savagely

BELOW_The Valkyrie has taken the art and science of aerodynamics to a level arguably never seen before on hypercars.

rapid on the track. The Valkyrie is centred around a carbon fibre structure and radical aerodynamics, which Aston states, "delivers unprecedented levels of downforce in a road-legal car, much of this is generated through underfloor aerodynamics". The hybrid car carries a 6.5-litre Cosworth V12 and an electric powertrain that offers a combined 1130bhp+ output. The hybrid battery system built by Rimac is said to be pioneering.

The lightweight construction has created a car with a power-to-weight ratio of greater than 1:1 (a feat also achieved by the Koenigsegg One:1). Only 150 road cars will be made, with a further 25 track specials, with rumours at the time of writing of a cost price of around £3,000,000 ($4,270,000). All Valkyries were sold before a single car was assembled. At the time, Newey told the assembled media that, "I've always been adamant that the AM-RB 001 should be a true road car that's also capable of extreme performance on track, and this means it really has to be a car of two characters. That's the secret we're trying to put into this car – the technology that allows it to be docile and comfortable, but with immense outright capabilities."

Arguably, never before has a supercar had this level of aerodynamic expertise and precision applied to its creation. The stylistic extremes of the car are a clear indication of the ultra-attention to aerodynamic detail. The teardrop-shaped cockpit

almost floats above the vast spaces that are cut underneath the bodywork. The huge full-length Venturi tunnels run either side of the cockpit door, making the Valkyrie unique in appearance and instantly recognizable. The tunnels draw in massive amounts of air under the body, which feeds the rear diffuser and therefore generates the ludicrous amounts of downforce on offer (one report suggests in excess of 1,800kg/4,000lbs at high speed). The upper shell is noticeably devoid of spoilers and trinkets, which avoids ruining the purity of styling but also, again, was done with pure aero in mind.

"All Valkyries were sold before a single car was assembled."

The low centre of gravity is achieved by, among other elements, the near-horizontal driving position, akin to a Formula 1 car. Inside the so-called glasshouse design shell, the cabin is also more like that of an F1 racer. However, it is Newey's near-obsessional pursuit of aero perfection that has dictated many changes to the exterior. This has produced what Aston have called "a genuine case of form following function". For example, changes have included openings in the body surface between the cockpit and front-wheel arches, after Newey found they were key to achieving considerable gains in front downforce. Even the wheels have been microscopically designed to aid airflow. Away from the aero, the car's weight

Hybrid tech uses a combination of an internal combustion engine paired with a variety of electric propulsion systems. This is not a science that supercars can lay claim to having invented. Frequently, as will be discussed on page 208–215, the science of supercars trickles down to road cars, just as race tech likewise often filters into supercars. However, in the case of hybrid tech, the process was inverted, as the initial ripples of innovation emanated from the lower end of the car market, with cars such as the Honda Prius, Civic Hybrid and Ford Fusion Hybrid being very early examples.

Aside from the first mass-market hybrids, there were a few supercar marques attempting to pioneer hybrid tech for this high-performance sector. An obvious example would be the 2011 Fisker Karma plug-in hybrid, and the early work of Tesla is also hugely important – before that company became the global behemoth of car manufacture it now is. However, it was not until 2013 that the science of alternative propulsion methods arrived in the world of supercars and it did so with the launch of three competing vehicles: the so-called "Holy Trinity" of the Porsche 918, the McLaren P1 and the Ferrari LaFerrari.

All three cars in the Holy Trinity shared very similar 0–60mph times and top speeds, although fuel consumption varied to some degree (see page 176). It was instantly clear that electric power cleverly coupled with internal combustion engines can deliver incredible additional torque. However, each car in this trinity offered very different overall characteristics and driving experiences.

> **"Alternative propulsion methods arrived with the Porsche 918, the McLaren P1 and the Ferrari LaFerrari."**

ABOVE AND OPPOSITE_The McLaren P1 was joined in the Holy Trinity by the Ferrari LaFerrari and the Porsche 918. All three used hybrid propulsion in different ways, but their combined impact made the supercar and indeed wider world reassess the potential of such technology.

HOW THE HOLY TRINITY APPROACHED THE SAME CHALLENGE DIFFERENTLY

What is the challenge? To incorporate hybrid technology as a technology demonstrator, but also to bring improved efficiency and emissions, plus increased performance. The cars showcase F1 and LMP1 racing technology on the road, but each has a different approach to charging and deploying electric motor and battery systems. (Across a number of track tests and lap times, the Holy Trinity appear to be within 1 percent of each other.)

Performance	Porsche 918	McLaren P1	LaFerrari
Active Aero	✓	✓	✓
Energy Recovery	✓	✓	✗
Electric Motors	✓	✓	✓
Twin Clutch 7 Speed Gearbox	✓	✓	✓
Carbon Fibre Chassis	✓	✓	✓
Turbocharged	✗	✓	✗
Four-wheel steer	✓	✗	✗
All-wheel drive	✓	✗	✗
Carbon Ceramic Brakes	✓	✓	✓
Electronic Adaptive Dampers	✓	✓	✓
Weight-to-Power (lbs per hp)	4.26	3.42	3.18
0-60 mph (seconds)	2.5	2.8	2.9
Cubic Capacity (Litres)	4.6	3.8	6.3
Total Power (bhp)	880	903	970
Electric Motor Power (bhp)	280	176	170

MCLAREN incorporate their Kinetic Energy Recovery System (KERS) motor in the engine block and recover braking (and coasting) energy, charging a battery pack in the chassis. McLaren place value on their active aero and suspension systems. The roll stiffness in race mode increases by more than three times, the ride height lowers by 50mm (2in) with underbody and diffuser plus rear wing hustling as much low-pressure air as possible, creating huge suction in the corners and backing off into a more drag-sympathetic mode on the straight, including drag reduction system.

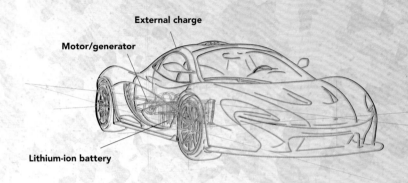

External charge

Motor/generator

Lithium-ion battery

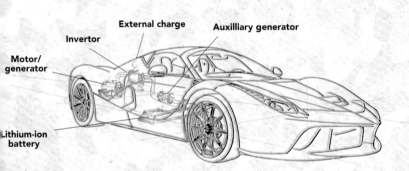

External charge

Invertor

Auxilliary generator

Motor/generator

Lithium-ion battery

FERRARI use HY-KERS, a system that maintains charge in a battery in front of the engine. The Motor/Generator Unit (MGU) at the back is driven by one of the two primary shafts in harvesting mode. When the engine calls for power, the invertor switches and the MGU goes from harvesting (generator) to a motor and transmits drive through the same primary shafts in harmony with the Internal Combustion Engine (ICE). V12 high compression and high revving (9,500rpm) muscle mixes with the HY-KERS (unlike the P1 and 918, the HY-KERS is engaged at all times and not used as a boost). The LaFerrari has no full electric mode.

PORSCHE deploy more electronics and controls with additional systems (such as four-wheel steer, all-wheel drive) through the drivetrain, suspension and steering with 30 percent more electric motor power, which justifies the additional weight of these systems (endorsed by the 0–60mph time). An electric motor for the front and the rear axle with energy recovery through engine braking and inverted for electric boost. Like the P1, the 918 has a full electric drive mode for limited town driving and is recharged by the ICE.

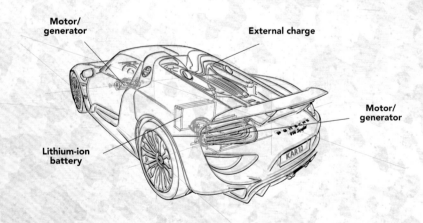

Motor/generator

External charge

Motor/generator

Lithium-ion battery

The Porsche 918 was the sleekest of the three. It offered rear-wheel steer of up to three degrees, which made cornering savagely accurate. Porsche have always been ardent adopters of new technology (from Ferdinand Porsche's first electric car through to the Panamera SE Hybrid). The 918 was conceived as the next generation Carrera GT and actually started life as a non-hybrid, but as attitudes and science within the automotive world changed, the head of the project, Dr Walliser, started to create something very special. In a previous interview with one of the authors, he said, "Until that point it was not clear if supercars could even survive in the future, if they would get social acceptance, what about fuel consumption, regulations, homologations and so on? It was unclear if there will be a next generation of super-sports cars at all. Perhaps they will just die, the dinosaurs of automobiles? ... A super-sports car must be relevant for its time. We must strive to be pioneers." Porsche were able to utilize the electric drivetrain in their car, add a luxurious cabin including a highly innovative entertainment system and still bring in the fastest ever time (at that point) around the Nürburgring (6 minutes 57 seconds) with the car in automatic and the suspension set to soft.

"Porsche have always been ardent adopters of new technology."

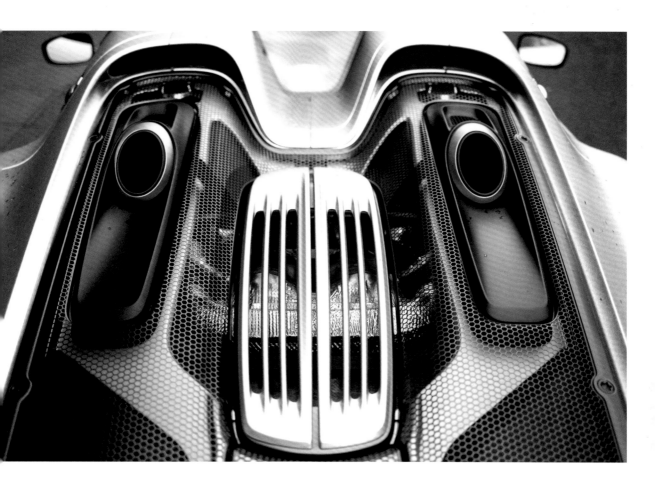

Magnus Walker, aka the Urban Outlaw, is a world-renowned collector of Porsches and is very impressed by the 918: "Your grandma can go drive it to get groceries in regular non-sport mode. However, you turn the button to Sport or Sport Plus and all of a sudden, it's a streetable race car that can do under seven minutes at the Nürburgring. That's what the science of these modern supercars offers – the diversity of use."

The McLaren P1 is arguably much less civilized and that is meant as a compliment. It is a purebred race car for the road, yet it is still useable and some customers are reported to drive theirs every day. Following on from the F1 was a massive challenge but the P1 did that with aplomb. McLaren Automotive test driver Chris Goodwin says: "We are selling the car to a whole range of drivers, not just world champions or experts, plus they are all driving in different environments. It's very easy to build a racing car that's going to be driven by professional racing drivers on smooth race tracks. But what about out there, on normal, bumpy roads? So we had to build a car with a whole different range of characteristics and the only way we were able to do that was to introduce a lot of technology. To do that takes immense attention to detail. There are a thousand aspects that we need to pay absolute attention to. All these things have to be evolved in minute detail."

That technology included active aero that was derived from the marque's F1 team and was unique to the car. The rear wing creates so much downforce that if it was not retracted at certain points it would break the rear suspension. The weight-saving obsession extended to not using lacquer on the carbon fibre to snip 1.5kg (3.3lb) in weight and using glass that was 1.5mm (0.06in) thinner, complete with a chassis weighing less than an average adult male. The wheels are military-grade aluminium. The car uses silicon carbide, the hardest substance known to man.

Ferrari LaFerrari obviously introduced Italian styling and verve into the hybrid mix. However, it was also a scientific tour de force, too. Boasting the most power of the three Trinity cars, as well as the lightest kerb weight, on its launch it was the quickest Ferrari ever built.

All three Holy Trinity cars quickly sold out and pre-owned examples were changing hands on the open market for hugely inflated re-sale prices. Therefore, their commercial validity is not in question. Their differing technological approaches to the same propulsion challenge is also fascinating and has certainly added to the science of the supercar. However, the most significant impact of the Holy Trinity in the supercar world may actually be generated by something other than the brilliant science on offer – what those three cars did was convince the supercar buying market that hybrid technology was not only a viable option, but potentially a superior one to the old guard. Also, the way that all three cars took science that was intended to reduce emissions and instead subverted it for massive performance gains was decisive. When supercar buyers and fans all over the world saw the statistics of these three machines, hybrid tech was accepted by the vast majority almost overnight. Intentionally or not, the Holy Trinity acted as hybrid technology's greatest ever PR coup. When the science in these three landmark cars is superseded or becomes outdated – which is almost inevitable – their ability to convince the car world as well as wider society of the validity of hybrid technology will be the Trinity's greatest legacy.

"Intentionally or not, the Holy Trinity acted as hybrid technology's greatest ever PR coup."

In the wake of the Holy Trinity, a number of other supercar manufacturers have pioneered their own approaches to hybrid tech. Koenigsegg is certainly one of the most adventurous and unconventional of these companies. Previous models had trialled the use of biofuel (also known as flex-fuel) and innovations such as dihedral doors, but most notably they had quickly established a reputation for blistering performance. "The philosophy at Koenigsegg is one focused solely on performance," says the company's founder, Christian von Koenigsegg. "I like to invent things. I thought that one of the differentiating factors that could make my creations desirable to the sports car audience was to make them innovative, more exciting and basically avoid compromise as much as possible." Koenigsegg's cars are

dripping with high-tech, but for the purposes of this chapter, consider the staggeringly fast Regera (Swedish for "to reign"), launched in 2016 as "a luxury Megacar alternative to Koenigsegg's traditional extreme, light weight, race-like road cars". The car's performance statistics are breath-taking: the dry-sump twin-turbo five-litre V8 engine is combined with three electric motors and pioneering battery power, with total horsepower in the region of 1,500bhp on offer. Arguably the most innovative part of this car is that it uses a single-speed fixed-gear transmission, known as the Koenigsegg Direct Drive System, instead of a conventional gearbox. This is due to Koenigsegg's belief that traditional hybrids represented a compromise in terms of weight, complexity, cost, packaging and efficiency.

"I had been pondering and thinking about making some kind of electrified car in combination with our extreme power and weight dense combustion engine," says Christian von

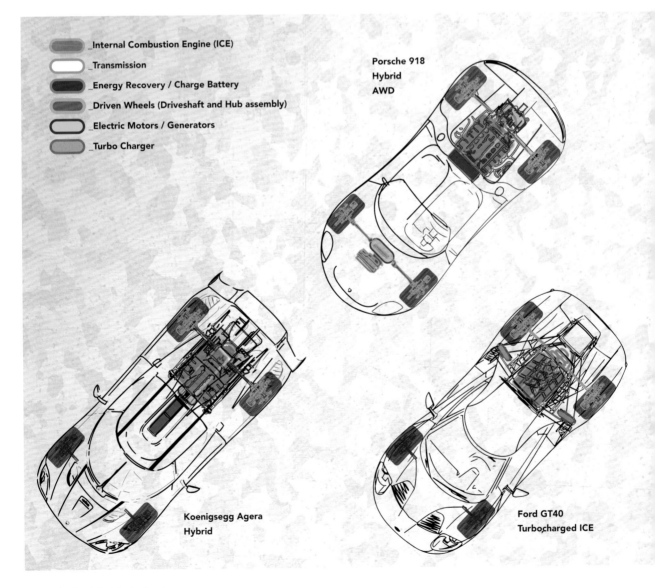

_Internal Combustion Engine (ICE)

_Transmission

_Energy Recovery / Charge Battery

_Driven Wheels (Driveshaft and Hub assembly)

_Electric Motors / Generators

_Turbo Charger

Porsche 918
Hybrid
AWD

Koenigsegg Agera
Hybrid

Ford GT40
Turbocharged ICE

Koenigsegg, "but it can all fall apart on the weight/size issue of electric motors and batteries. However, then we came up with the idea of removing the transmission completely, taking out one of the heavy and most complex parts of the car and having direct drive from the combustion engine with the help of an electric motor – suddenly the compromise was much, much less. For example, you have the lowest energy losses you can, because you have no transmission losses. That's what enabled and kicked off the programme, that idea to avoid compromise again. To have all the upsides but remove as many of the downsides as possible. In my mind it's the ultimate combination of electrical propulsion and combustion propulsion you can have, regardless of the type of vehicle."

Hybrid developments could fill a book in their own right, but below is a selection of the ingenious ideas currently swimming around in the minds of the supercar designers across the globe.

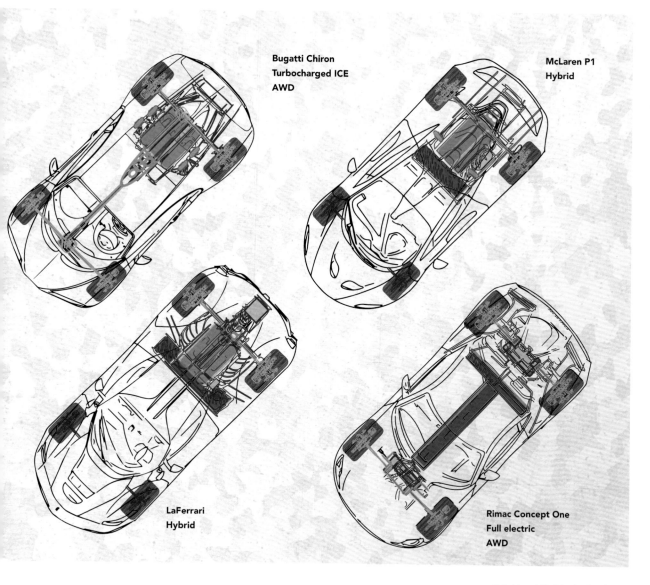

Bugatti Chiron
Turbocharged ICE
AWD

McLaren P1
Hybrid

LaFerrari
Hybrid

Rimac Concept One
Full electric
AWD

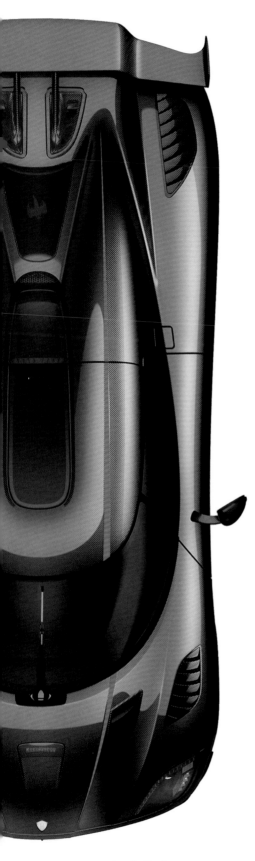

Christian von Koenigsegg is at the very forefront of this revolutionary science and has a fascinating perspective on what the future may hold for hybrid tech: "In the short and the long term, it's a slightly different picture. We are seeing great strides and opportunities in the combustion engine arena. For the short term, electrical batteries are so heavy that they can't compete on a power-to-weight ratio. I think we will see hybrids going from heavily combustion-engine oriented to 50–50, then maybe more electrified as the batteries grow smaller and lighter."

There are clearly still significant drawbacks and challenges for this new science: the sound of some of these cars is less than inspiring. Jumping in a high-performance car and turning on its savage engine is one of the most exciting and rewarding benefits of owning such a vehicle. Pulling away in silent electric mode, however, is much less invigorating. Some supercar manufacturers are now toying with artificial sounds to make the experience more adrenalized, much to the horror of many purists. There have been unsubstantiated rumours of some owners of hybrid supercars asking certain marques to remove the hybrid technology altogether (with one disappointed owner calling his hybrid hypercar "an industrial-strength electric kart"). Also, the range anxiety issues of city cars are multiplied many times with a car that can shift through the miles so quickly.

Peter Read, chairman of the Motoring Committee and board member of the Royal Automobile Club, has his own personal reservations: "All the major manufacturers of supercars are jumping on the bandwagon of hybrids with or without turbocharged internal combustion engines or pure electric, but there are some marques who are suggesting their top-end cars will continue to be powered by a naturally aspirated, classical V12 engine. Possibly they may use a V6 or V8 turbo-charged unit for their entry-level models and even have a hybrid in there somewhere. Of course, this is all subject to the rapidly changing regulations."

Read has a penchant for a naturally aspirated V12 – that is for him the defining science of any supercar: "While manufacturers might get a V12 through the emissions police because of low production numbers, they might come up against local governments desperate to appease the tree-huggers by banning such engines in towns and cities. Eventually, although they can build normally aspirated V12s, which sound fabulous, the owners might not be able to use them anywhere."

Read continues, "A second issue is, what happens when all this new technology fails? That's fine if the ECU [Engine Control Unit] in your hybrid supercar is still current, but what happens if your supercar is ten years old and they no longer make those ECUs? It's not a simple case of 'take one computer out and replace it with the latest upgraded one' – these cars have 40 or 50-plus computers, which all need to talk to one another. Will an upgraded fuel-injection ECU still be able to communicate with an older-generation gearshift one and, if not, does everything need

replacing? If so, where will you get those spares? The basic V12 supercar on carburettors and a distributor can always be fixed, so a Bentley Blower can go on forever, but a LaFerrari ... who knows?"

Bugatti appear to have forseen such future-proofing issues and have built their latest car, the Chiron, with expansion ports for future iterations of processors. This should, in theory, allow the car to remain compatible with the ever-developing peripheral hardware and software for many years.

Gordon Murray also has some intriguing questions about these hybrid cars: "I would like to ask them why they are doing it, because they don't need it for fleet emission packages, because they are only selling a few hundred of these things. So is there a genuine interest in developing new technologies and handing that down the line?" Murray also has little patience with supercar companies who complain that legislation is stifling their work: "That world is still far less restrictive than designing a Formula 1 car, which got quite ridiculous. Relative to what F1 can be like, regulations for road cars are not stifling at all." Gordon also shares the same concerns as Peter Read: "If you want the supercar to one day become a classic and an icon, trends are very dodgy."

The above reservations about the constant march of automotive tech are valid concerns, but the supercar world

> **"I think we will see hybrids going from heavily combustion-engine oriented to 50–50, then maybe more electrified as the batteries grow smaller and lighter."**
>
> Christian Koenigsegg

ABOVE AND OPPOSITE_ Koenigsegg's ascent to the upper echelons of supercar performance tables is due to the marque's highly innovative and unique approach to age-old issues of supercar science.

cannot turn back now. The regulations weighing down on every motor car are becoming ever more stringent – in 2017 the UK government announced that purely petrol and diesel cars and vans will be banned from 2040. If this stance is replicated around the world, the car industry is headed for an entirely new dawn. Hybrid electric tech may not even be the ultimate winner – no one has a crystal ball – and other methods have also been experimented with, such as using hydrogen, for example (at the time of writing, both the infrastructure and public confidence in this alternative fuel are limited, even though you could in theory drink the exhaust emissions – pure water). Therefore hybrid tech has to evolve at quite a pace and find a legitimate and practical role for all cars, not just super ones.

There are other related challenges ahead. The surge towards full automation seems relentless and although at the time of writing that science is still in its relative infancy, companies like Tesla have cars out on roads in the USA that are using this technology. There are many difficult issues still to be resolved, such as liability in the event of an accident, as well as privacy and hacking concerns, but these apply to all road cars, not just supercars. It seems likely that even when self-driving cars are commonplace on the roads, the supercar may well remain the exception. After all, why buy a stunningly engineered, state-of-the-art driver's machine with the greatest performance money can buy and then sit in the passenger seat?

Supercars demand a different motive. Supercar owners are looking for *more* of a driving experience, not less. If a car is designed to do anything other than simply go from A to B, then clearly speed, enjoyment of the drive and performance have become factors in its creation. You can buy a phenomenally economical family car for a few thousand pounds, with luggage space and all the options; or you can buy a hypercar for several million pounds that will get you to the same destination faster and in considerably more style. But you still end up at the same destination. So the *experience* has to be the difference.

Self-driving cars and supercars are never going to be perfect bedfellows. Whether that paradox ultimately extinguishes supercars altogether remains to be seen, but regardless of how brilliant self-driving technology becomes, for millions of people around the globe the very notion is anathema to the attraction of cars in the first place. You can develop and evolve science, you can change how cars are built and how they perform, but you can't change the human instinct to enjoy driving.

In summary, Nick Wirth of Wirth Research has a very positive and energized view of the way supercar technology (and everyday car science) is evolving: "Manufacturers are engaging in the next big thing right now. There is revolution in the air."

> **"Hybrid tech has to evolve at quite a pace and find a legitimate and practical role for all cars, not just super ones."**

CHAPTER 12

HOW TO DESIGN A SUPERCAR

The history of supercar science in this book is necessarily presented in neatly defined chapters and sections. However, the business of actually designing a new supercar has far less defineable parameters.

OPPOSITE_Named after a pre-World War II racing hero, the Bugatti Chiron had the massive task of superseding a supercar legend.

With the expectations for the capabilities of any new supercar constantly increasing, designers have to take on board the science, technology, history and provenance of both the entire supercar world and also their own particular marque before they can begin what is a hugely complex and arduous process, littered with thousands of crucial decisions.

This challenge is even greater if you are trying to improve on the fastest production car ever built, the Bugatti Veyron. That is exactly what Achim Anscheidt, the director of design of the Bugatti Chiron, was faced with when he started to create the Veyron's successor. In the Orangery at Bugatti's refined headquarters in Molsheim, in the Alsace region of France, Anscheidt walked one of the authors through the complex and at times ultra-demanding process of designing what may well prove to be the greatest internal combustion engine car the world will ever see. In the process, he sheds a hugely insightful light on the world of supercar science that has challenged generations of engineers and designers before him. His revelations detail the enormous task facing any supercar designer, but also symbolize all the various elements of supercar science that this book has highlighted, all of which need to be considered when developing a new high-performance vehicle.

RIGHT_ Bugatti Chiron, 2016

ABOVE AND OPPOSITE_Every element of the Chiron was designed up to, and in some cases potentially beyond, the known levels of supercar science.

The first challenge when developing any supercar is bettering the decades of engineering triumphs that have gone before. "I must tell you honestly," recalls Achim Anscheidt, "I had sleepless nights, how to go about this huge challenge, to create this next big step. A new modern-day Bugatti is not created every two or three years, this is not a small task. I just didn't want it all to go wrong. It was not like from the very first moment I knew exactly how this should play out."

Bugatti's president, Wolfgang Dürheimer, agrees: "I think the biggest challenge is to make every component, every item, every criteria that the car delivers remarkably better than the previous car so that in a 20-minute test drive it is immediately apparent that this is a completely different proposition to the predecessor."

"The value of a Chiron is undeniably in the engine."

Achim Anscheidt, chief designer of the Bugatti Chiron

The first engineering task facing a supercar designer is the engine, the power plant that began this whole story back in 1885 with the Motorwagen and that is only now beginning to evolve towards a different ethos. For the Chiron, the crucial decision was made to *not* use hybrid technology. Instead, the Chiron has the world's first production sports car engine with 1,500bhp. Eyebrows were raised when Bugatti chose not to enter the hybrid race, but there was a very pure logic behind this decision, as Anscheidt explains: "The beauty is this W16 engine concept is tough to copy. This has always saved us over all these years, helped us to stay unique. We have been copied in price and a lot of other elements but nobody can copy that engine because it is such a huge project to develop. The Chiron therefore is the last, most brutal rendition of a pure internal combustion engine. I believe you can already see that it was the right decision."

"The value of a Chiron," he continues, "is undeniably in the engine. It is the drivetrain, it is the power plant. You can talk about design as much as you want to. However, if there wasn't this power plant, then it would be not worth the money and the talk. The engine makes the product." What the Bugatti engineering department created with the Chiron's engine will surely be regarded as arguably one of, if not *the* most advanced iteration of the internal combustion engine.

Even so, such a massive engine requires huge supporting tech and in the case of the Chiron this illustrates how supercar

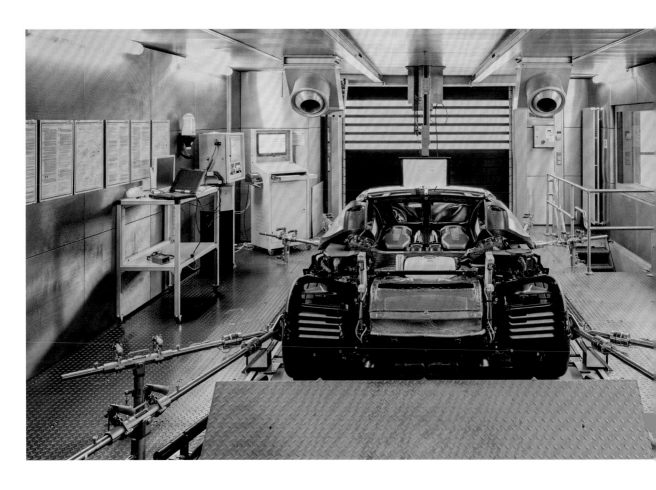

manufacturers often have to invent new manufacturing facilities and methods to keep up with their own science – just as the Jaguar team had to with the revolutionary E-Type back in 1961. The Bugatti dynamometer for testing the power plant had to be redesigned and built especially for the Chiron, as existing facilities were inadequate to simulate the loads created by this enormous new engine. Although the legendary W16 engine from the Veyron provided the basis for the new car, it was almost completely redesigned, creating a 1,500bhp output and, perhaps even more significantly, 1,600Nm torque at as low as 1,800rpm. This means the car's apocalyptic performance is on tap very much earlier and far more savagely. This was achieved by the use of a pioneering two-stage turbo set-up, which uses four very large turbos (69 percent larger than on the Veyron), with two always spinning. This creates a more sequential and smooth burst of power, such that turbo lag is effectively rendered obsolete by this system. The Chiron moves off the mark with only two turbochargers in operation, then the other two units are activated at about 3,800rpm, which delivers an absolutely linear power curve from 2,000rpm and huge torque in the low engine speed range. In layman's terms, this means that the huge

"For any supercar, it is not enough to just have brute power."

maximum torque of the Chiron is available for over 70 percent of the entire engine speed range, as Anscheidt explains: "The most underrated thing about both Bugatti cars is the torque. With double stage turbos you have maximum torque at 1,800rpm, offering maximum performance with outstanding control in all speed ranges." Feeding this monster is made possible with the help of a duplex fuel injection system with 32 injectors, essentially a direct descendant of the initial principle of fuel injection pioneered way back in the fifties by the Mercedes Gullwing.

For any supercar, it is not enough to just have brute power. Weight is an issue, as seen with the use of composites dating back to cars such as the Ferrari 288 GTO and F40, plus of course the McLaren F1 and the Pagani Zonda. In the Chiron, extensive use of lightweight materials such as carbon fibre and titanium are everywhere, including a custom exhaust system and a lighter crankshaft. Even the intake tube, the charge air system and the chain housing are all made of carbon fibre (with an integrated rubber joint). Other innovations include a titanium silencer, six tail-pipe exhaust and six catalytic converters whose active surface area would cover 30 soccer pitches.

Simple science tells us that if power cannot be applied to the road, then no supercar will go anywhere. Therefore, as shockingly powerful as the Chiron's engine is, handling all this power required the engineering team to upgrade the revered DSG gearbox in the Veyron to one that has the largest, highest-performance clutch ever fitted to a passenger car, all assisted

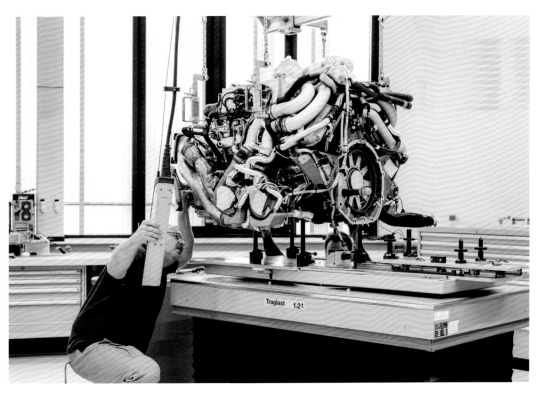

OPPOSITE_
Enormous
research and
development
was expended
analysing how air
entered the Chiron
and how it could
be manipulated
to help the vast
cooling challenges
generated by the
sheer power of
the car.

by permanent four-wheel drive. Transferring a supercar's power from the engine to the gearbox and down on to the road is even more crucial – the Chiron's tyres must be capable of transferring a maximum torque of up to 5,000Nm per wheel safely to the road with unprecedentedly high lateral guidance. Consequently the tyres had to be tested on aerospace industry test rigs.

Many years before the Chiron, Ettore Bugatti overheard someone criticizing his car's brakes, and replied, "My cars are made to go, not stop!" As entertaining as this answer is, the Chiron required sledgehammer brakes to pull the beast back into line from vast speeds. The special brake discs made from carbon silicon carbide (CSiC) are lighter and less liable to corrode, and hugely powerful for stopping this machine.

As with the Gullwing, Miura and other high-performance cars with mid-engined layouts and lightweight chassis, the Chiron's chassis had to be eminently capable. As with the 959, electronic assistance is utilized to perfect the various ride modes on offer. The massive speed and acceleration requires an immense level of stability and so, for the first time, Bugatti used an adaptive chassis with five driving programmes for greater agility and ride comfort (which, by definition, also offered supreme safety benefits). The highly sophisticated carbon fibre monocoque offers extremely high rigidity. Frequently the so-called sandwich construction is used in the composites, such as on the underfloor, a method of composite construction championed by Gordon Murray, the creator of the McLaren F1.

Just as with all powerful supercar engines that generate large quantities of power, Anscheidt and his team were next faced with arguably the biggest scientific challenge – cooling. Early tests saw the rear end of the test cars melting from the heat generated. "Percentage wise, the heat rejection increase from the Veyron to the Chiron was in the region of 50 percent," Anscheidt explains. "The power plant generates over 3,000bhp of heat that has to somehow be dissipated. That was a major headache." Therefore the cooling systems, including ten radiators, had to be prodigious. The coolant pump alone is capable of circulating 800 litres (176 gallons) of water through the engine in one minute. This is helped by the 60,000 litres (13,200 galllons) of air per minute that are pumped through the engine.

All supercar designers and developers are faced with massive scientific challenges, but they are not always armed with an endless amount of time or money. "We had this thermal issue in the rear of the car with the air stream that we needed to create in order to get the hot exhaust gas further behind the car, but you don't have a lot of time to find the answer," says Wolfgang Dürheimer, Bugatti's president. "You need to find a solution, it needs to work and it needs to come despite hell or high water."

A crucial tool for Anscheidt to handle cooling was aero. As seen from the early land speed cars such as Jenatzy's "La Jamais Contente" right up to the present day with the hyper-advanced aero of the Aston Martin Red Bull Valkyrie, aerodynamics can make or break a supercar's performance. For the Chiron, both wind tunnels and CFD were used extensively (more than 300 hours in the wind tunnel), a dual approach favoured by Wolfgang Dürheimer: "I never believed that wind tunnels were finished. We always did a lot of work in the wind tunnel and it's also the difference of real life and simulation. When you look at the rear end of the Chiron it is like a stealth fighter, but this is not for looks, it has a very specific aero and cooling function."

No modern-day supercar team can attempt to improve the genre without having a huge knowledge of both the science and the history of high performance cars. At the same time, for all the glitz and glamour of a supercar launch, the team behind that vehicle will no doubt have endured many moments of extreme stress when it appeared the science just did not work. Having an over-arching knowledge of all these areas of design is what equipped Achim Anscheidt to rise to such a daunting challenge. A coherent, overall approach to aero was the lightbulb moment – he suddenly realized what he needed to do to achieve the ultimate with this new supercar. The staggering power created monumental cooling challenges that initially had an almost suffocating impact on Anscheidt's capacious design brain: "I had hundreds and hundreds of drawings and also dozens of models made up in the studio, but something wasn't right. I said, 'This is not going the right way'. What I had was just stylistic play but that didn't do the project justice because styling of itself is not valuable enough, it is just a designer moving a line up and down, it cannot be just that. Shape is not enough.

"Then slowly but surely it dawned on me what was required. The engine department was delivering the most incredible 1,500bhp power plant and I thought, 'My God, this is a monster'. Impressive certainly, but the demands that placed on *every single component* and *every single decision* in the process was massive, so severe. That is why it dawned on me how we had to do this; there was only one genuinely authentic and true path – to tailor all aspects of the car around *performance*. That had to be the over-riding philosophy, 'Form Follows Performance'. That way I don't have any excuse to draw any line as a stylistic, fashionable statement, I can let these performance elements do all the talking. This helped maintain the integrity of my design team, everything we did from then onwards had a purpose and a purity."

> ## "There was only one genuinely authentic and true path – to tailor all aspects of the car around performance."
>
> Achim Anscheidt, chief designer of the Bugatti Chiron

Dürheimer describes it thus: "I call it the concept harmony of the entire project; it needs to appeal to your visual senses, the car needs to look stunning but it still needs to have the functionality. Achim and his designers have a deep understanding about technical issues and were able to create a car with pure functionality, and to *package* it so that the entire concept also looks cool."

Just as the designers of the Jaguar E-Type used aero to create a slippery car but also a beautiful one, then so too were

OPPOSITE_Sketches and real-life works of art.

RIGHT_Such is the heat generated by the exhausts that early mule cars saw their rear-ends simply melting when tested.

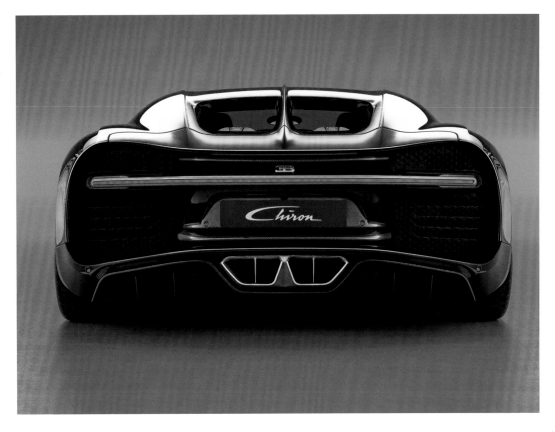

OPPOSITE_The central spine mimics the iconic Bugatti Atlantic while the open engine bay is needed to assist with cooling such an enormous power plant.

the proportions of the Chiron bound by what was necessary as a performance priority. This was not an easy task. For example, the architecture at the front of the car, with the famous horseshoe grille and distinctive eight-eye headlights, complement flowing lines that actually provide ingenius air intakes. "That design is not just a fake or for any kind of show or stylistic detail," Anscheidt explains. "It is helping us on the cooling development of the brakes, an important aerodynamic function." This intelligent air intake management significantly improves aerodynamic properties as well as enhancing vehicle cooling, all coherently designed by Anscheidt to look sensational.

For the Chiron, the so-called C-bar that swipes around the side of the car (also known as the Bugatti line) is another example of a style detail that is far from just aesthetic. It allowed Anscheidt to direct huge amounts of air along the side of the car and into the engine compartment at various key points (crucially higher up, creating better uninterrupted airflow). The aerodynamically designed rear quarter panels also serve the purpose of creating a beautiful reflection. "I like to say to my younger designers that simplicity is a precious element of complex refinement, by which I mean to successfully be ultra-simple about your shapes and design is sometimes very hard," Anscheidt explains. "The things that are there need to be used for a reason. So with this in mind, how

"The rear quarter panels create a beautiful reflection."

ABOVE_The Chiron
uses design cues
from traditional
Bugatti cars
and melds this
with the very
latest scientific
innovations.

do we judge quality? It's by a reflection. When you look at a shiny object you look at the reflection and if there is a problem you see the fault in the surface. This is why the reflection travels with so much grace, acceleration and deceleration on that part of the car."

The exterior styling also brings us to the point that all supercar designers face – maintaining the pedigree and DNA of the brand without looking like a repetition of old glories. How do you design a Ferrari that does justice to that great marque? How do you create a new McLaren hypercar when the F1 is still so universally adored? The accentuated lines of the Chiron were inspired by the legendary Bugatti Type 57SC Atlantic – most obviously the sinuous spine flowing over the Chiron, which mimics very closely the ridge on the Type 57SC (a feature that Piëch on the VW board was adamant they had to keep, as he is an ardent fan of that older car).

Other scientific elements culled from decades of supercar science and honed to new extremes for the Chiron include underbody NACA ducts to assist with cooling, an air curtain, an aerodynamic front splitter, air intakes for the oil cooler and engine air inlets on the sides as well as the enormous air-brake itself (which alone can generate 240kgs (529lbs) of downforce under braking).

As noted, gone are the days when a supercar can be cramped, badly fitted inside with poor visibility and inferior ergonomics. Bugatti as a brand epitomizes luxury so with this

history in mind, the interior of the car is sumptuous, with better headroom, a high-end stereo, remarkable soundproofing and a central console carrying only the most essential controls. The first airbag in the world to shoot through a carbon fibre housing helps safety, while there is even room for a bespoke suitcase trolley approved for air travel. All supercar designers have to combine safety, quality and comfort within a cabin, using the latest science at their disposal; like Murray before them with the F1, Bugatti strive for *useability*.

Learning lessons from the best and worst of electronic gadgetry, the Chiron utilizes both new and old school. The instrument panel offers two digital read-outs either side of a mechanical speedometer. "The speedometer is the most vital and important instrument on a modern-day Bugatti," Anscheidt explains. "The idea that a young lad can peer into the Chiron's window and see 500kmh [310mph] was very important to us; if you have just digital read-outs only, then when the car is parked it is all blank, dark and has no emotion. Our speedometer is a small, mechanical analogue wonderworld."

Wolfgang Dürheimer, Bugatti's president, points out the level to which the attention to detail reached: "We followed pretty much the ethos of Ettore Bugatti that nothing is too precious, nothing is too exclusive to not end up in a Bugatti. So we have the one-carat diamond speaker membrane of the sound system in the car. Some people have diamonds on their necklace and

ABOVE_Bugatti cannot afford to create a car that has performance but no luxury – the art of designing a supercar in the 21st century recognizes that many customers demand both.

ABOVE_The
speedometer, with
a potential top
speed of 500kmh
(310mph) proudly
emblazoned on an
analogue dial.

some have them in their car speaker, to give the sound system an even more crisp sound."

Like the McLaren F1 before it, the Chiron was a complete and articulate package. Anscheidt insisted on a consistency between the exterior and interior. "There is an over-arching theme – the C-shape on the exterior of the car often gets mistaken for the romance of Ettore Bugatti's signature or even the signature of Louis Chiron, but that's not the case, it's really an element that relates strongly to the interior. So, the exterior and the interior development have a strong architectural resemblance."

Decades of evolving wind tunnels, CFD, track runs and workshop rigs have proved that it is essential for any supercar to maximize the science available by conducting exhaustive development and testing. For the Chiron, the entire development process saw Anscheidt's team complete more than 650,000 test kilometres (about 404,400 miles) and use up more than 200 sets of tyres. As with electronically advanced cars such as the Ferrari Enzo, this entire masterpiece is managed by 50 computers and the assembly of over 1,800 parts takes six months per car, completed by a small team of just 20 engineers. Only 500 Chirons will be made and at the time of writing, 300 of the £2.1 million ($3 million) cars were already sold.

> **"Like the McLaren F1 before it, the Chiron was a complete and articulate package."**

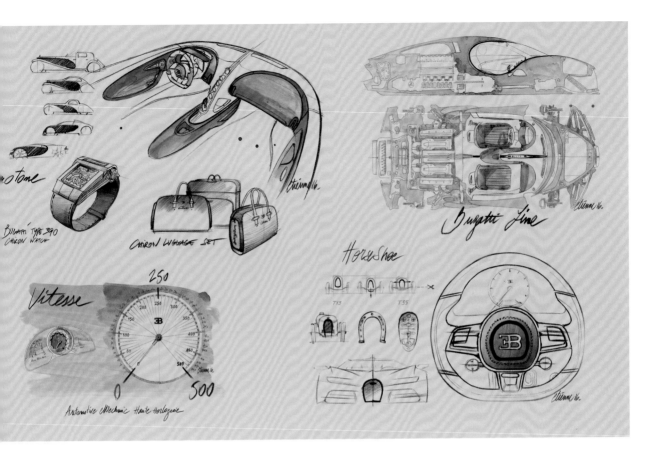

Monotone

Bugatti Type 390
Chiron Watch

Chiron Luggage Set

Bugatti Line

Horseshoe

T13   T35

Vitesse

250

150   200   250   300   350

100   3B   400

50   450

0   500   500

Automotive Mechanic Haute Horlogerie

When developing the Chiron, it is clear that Bugatti respected, absorbed and then advanced every element of supercar science that this book has highlighted. In so doing, the car represents an ideal summary and template for the extraordinary levels to which high performance car science now has to reach. It is clear that hundreds of lessons have been taken from the history of the science of supercars and applied to this new record-breaker at a level that is simply breath-taking. In so doing, what Bugatti have created is a car that is so superior to its predecessor, the Veyron, that it is actually quite hard to accurately put into words. As a symbol of the science of these supremely exotic cars, as well as the complex process that goes into designing every new supercar that comes to the market, there might just be no better example.

Whether the Bugatti Chiron is your favourite supercar is a highly subjective point. It is almost certainly the greatest internal-combustion-engined vehicle that will ever be made, given the ever-changing climate for alternative fuel sources. What Achim Anscheidt and Wolfgang Dürheimer have explained above is just a small measure of the colossal effort, thought, ingenuity and application that has to go into mastering the science of any modern supercar.

ABOVE_The design process for the Chiron was one of the most exhaustive and thorough ever conducted for a supercar.

CHAPTER 13
THE FUTURE

In the second decade of the 21st century, a curious reverse-evolution of supercar science has begun: supercar customers have increasingly wanted a return to "old" technologies (or even older cars).

OPPOSITE_The spectacular Mercedes Project 1 supercar.

In the classic market, this is reflected by manual cars being worth more than the so-called higher-tech paddle-shift versions that originally superseded them. "The manual car is not as good in many ways," argues Bugatti's Julius Kruta, "but you have the clack, clack as you feel the gear change, *you* operate the car. People have the feeling that they are deciding: you are less a passenger, you are more a driver. Yes, on paper these cars are slower, they are more complicated, you have the problem of the clutch, it's a nuisance! But people want their gearboxes again." Several manufacturers are even adding slight "clunks" into their double clutch gearbox systems to re-inject an element of the gear actually changing. This situation is not limited to just gearboxes – the array of electronic driver aids is also turning some people off. One of the authors knows of one particular supercar collector who sold his Enzo – a car that was lauded on launch for the dazzling electronic systems on-board – and bought his F40 back because he was "tired of feeling that there were so many computers between me and the road".

Another challenge with the rise of the science of supercars is the question of where you can safely and legally use such tremendous performance. The rumour is that the Bugatti Chiron can do a top speed comfortably 200mph more than the 70mph

RIGHT_Gordon Murray's new supercar was unveiled in late 2017 under his IGM brand. It combines lightweight aluminium with his groundbreaking iStream technology, which has revolutionized the previously laborious process of manufacturing carbon fibre.

speed limit in the UK, for example. It can surpass that same speed limit in first gear. Le Mans race driver Andy Wallace is professionally used to very high speed but even he is staggered by the performance now on offer from road-going supercars: "When I was hurtling down the straight in Le Mans, racing close to 400kmph [250mph], I never imagined that I would ever have that situation again when I stopped racing. The very first time I drove the original Veyron, I was absolutely blown away by how fast the car was. It was all under control and all luxurious, comfortable and refined but the actual numbers on the speedo … you look and think, 'How on Earth am I going that fast?' The Super Sport version and then the Chiron are faster than any race car I have ever driven. That is a shock because for decades, motor racing was a step ahead in terms of the force, a very big step actually. Yet at a time when supercars have started to get more and more horsepower, racing cars have conversely started to get less and less. Also, when you are in a race car, you are suited and booted, it's a very hostile environment, it's hot, it's vibrating and so on, so you don't expect to slip into a beautiful leather seat in your civilian clothes and, given the right circumstances – a runway or *Autobahn* – be able to accelerate that fast and reach those top speeds. You have to do a double-take, 'Is this for real?'"

These challenges present a real and genuine dilemma – the evolution of supercar science cannot be wound back, but

> **"Where can you safely and legally use such tremendous performance?"**

ABOVE_The Mercedes Project 1 is the latest in a long line of high-performance cars that started with the iconic Gullwing back in 1954.

there has to be a position where the raw essence of the driving experience is preserved. Dr Kerry Spackman is a renowned neuroscientist who has worked in motorsport for McLaren and Jaguar, and has a world-leading knowledge of how a driver's brain interacts with a car's performance: "Modern Supercars aren't just fast in a straight line (400kph/250mph), have blistering acceleration (0–100kmh/0–60mph in less than three seconds) or amazing cornering. There's something else about them that most people don't appreciate. With their lightweight composite construction, low rotational inertia and high downforce due to amazing aerodynamics, they're also *very* responsive. If you lose the rear end of one of these things without any electronic aids, it will happen far faster than the average driver can respond to. It's a bit like trying to return a tennis serve when someone is shooting bullets at you – it's just too fast.

"This is where the electronic aids come in. They can tame the car at the limit and step in and give the driver a 'hand'. The problem is, if these aids are too intrusive, the joy of balancing a car at the limit is lost – the entire visceral experience becomes benign. So there's a real balance between helping and removing the challenge.

"What makes this far more challenging at track limits is you don't want the 'driver's brain' fighting the electronic brain – which is really easy to do. The electronics react to the driver and the driver reacts to the electronics in a vicious circle. The aim of a REALLY good electronic system is to understand how this *particular* driver operates and what they're likely to do and then provide just the right input to make his or her corrections seem magic, without getting in the way or taking away the challenge. This is the brave new frontier of Artificial Intelligence aids – a combination of neuroscience, electronics and 'deep learning'."

Christian von Koenigsegg stresses that the realm of supercars cannot be just an exercise in science: "We are building a handful a year, so we are not going to pollute the planet given that most customers only use them every other month or whatever, so the emissions issue is irrelevant to a certain degree. Nonetheless, we have to comply with the regulations. But all that said, the emotional aspect is more important than the environment when it comes to these cars. From an *emotional* perspective, these cars are not only about efficiency or fuel consumption and the environment, they're also about enjoyment."

So what has over a century of supercar science given us? For every supercar detractor, there are a raft of technologies that these machines have bequeathed to the automotive industry for use on everyday cars. It is this developmental role, producing extremely pioneering science, which soon becomes commonplace, that is the greatest legacy of the science of supercars. This so-called "halo effect" or "trickle down" legacy is real and absolutely crucial to future supercars being funded.

For example, Koenigsegg's so-called direct drive system in the Regera hypercar is not just a supercar designer's self-

BELOW AND OPPOSITE_Christian von Koenigsegg revels in his unconventional approach to traditional engineering and design challenges and it is this maverick, individual approach that makes his supercars stand out as unique.

indulgence. This direct drive system will be applicable to large trucks but also city cars, so potentially Koenigsegg's tech could well have a massive filter-down legacy. He is not concerned if it meets with the approval of his supercar peers: "I am aware that many sports car companies would say removing the gears removes the fun part of driving and they would never want to do that, or even consider it. But we are not driven by what we

"Supercars have bequeathed a raft of technologies to the automotive industry."

should do from a historical perspective, we are driven from what we can do going forward to make the best possible product. So that gives us a completely different level of freedom. Since I'm the founder and the owner of the company, no one stops my craziness to do it!"

Although supercars may not have invented every safety feature found on a modern everyday car, it would be unkind to suggest the high-performance genre has not contributed heavily to safety – just because a car can be driven very fast does not mean the science behind its ability to corner prodigiously, brake with staggering force and protect its occupants in the event of an accident is not significant.

As Adrian Newey explains, much of the important influence of supercars on more normal vehicles' tech is defined by cost: "Supercars are the breed of road car that gets race car

technology first, simply because that particular technology tends to be very expensive and therefore the supercar owner can more easily afford it. Carbon composites are the most obvious example, but in a sense other exotic elements such as titanium are very difficult to be utilized by a mass-produced car and get to the point where that is cost effective.

"Nonetheless, on a more technical basis, looking at what science and technology can be brought forward into more general applications for mass-market cars is, for me, one of the most fascinating elements of looking into the future. How can these innovative ideas then be developed for more mainstream use in the future?"

You might wonder what single piece of science has been of the most significant benefit to the car industry – the DSG gearbox is certainly hugely important. That can be found in millions of family cars around the world. Similarly turbos are being used to reduce engine capacity and increase efficiency. Hybrid technology has certainly already had a major impact, but as previously noted, that was initially developed from the ground up in family cars. However, supercars have taken the baton and massively accelerated the evolution of hybrid science, so their role there is also important. Advanced electronic engine systems help with regulations and safety. Many everyday motors have benefited from engine tech taken from supercars and so on … the list is long and open to debate, including but not limited to other areas such as minituarization, advanced electronics, hydraulics, crash structures, aero advances and stylistic developments – all ideas found in everyday cars that were influenced by the laser-sharp focus of supercar designers.

> "Many everyday motors have benefited from engine tech taken from supercars."

However, Gordon Murray is in no doubt as to the greatest single gift from supercar science: "The use of composites. Initially that was very slow and very, very expensive technology. For 15 years I have been trying to take that expensive, slow technology and turn it into something affordable that you can do in under two minutes. So my company can take a panel that used to cost €600 and take five hours to create, but now we can make it for €20 in 100 seconds. In my lifetime, the single biggest transfer of technology from supercars (not all of them, I might add) is that technology, because it will allow everyday cars to be stronger, lighter, emissions will improve and so on. The crash tests alone are ridiculous – one series of tests we did for a major car company proved this technology was six times stronger than their steel cars."

BELOW_How the supercar is designed, manufactured, powered, steered, balanced and driven will all be open to huge debate in the coming years.

Bugatti's official historian, Julius Kruta, agrees with Murray about the significance of this material: "The use of composites in everyday cars will be the greatest impact from the supercar genre. That science has been phenomenal, and will only become more so. What interests me moving forward will be how are we going to recycle it in the future? Carbon fibre is a mixture of materials, it is a woven material, so it will be interesting to see that being recycled. That aside, its impact has been, and will continue to be, colossal."

Regarding the future use of composites and what might be the next leap forward for this science, Horacio Pagani has a studied opinion that concurs with Murray and Kruta's train of thought: "I think there won't be as radical a change as there was in the eighties and nineties, when we first began to use these materials instead of traditional ones. I think the most significant changes will relate to larger production runs, to increases in output, not simply using the materials on very special items such as supercars, but working on industrialization in order to produce a greater number of ordinary cars. This is something that's actually already happening ... I think the key is being able to produce at lower costs.

"I don't know when composite materials will become more competitive than traditional ones, perhaps in a few years, but I believe that's where the challenge is, not elsewhere. The material will undoubtedly evolve: for example, we created new materials for the Huayra Roadster and managed to increase their rigidity, the material's mechanical characteristics, by 52 percent, though admittedly we did so in an artisanal manufacturing context such as ours."

The science of supercars is an ever-expanding world of high innovation fuelled by competing technologies, commercial forces and sheer engineering and design genius. The reality is that no one knows which of these technologies will triumph; equally, it is not yet apparent which manufacturer will produce the next "game-changer" supercar or even what area of automotive

BELOW_The Science
of Supercars
– perhaps in
many ways the
journey is only just
beginning.

science it will advance. At the time of writing, for instance,
3D printing of parts is potentially offering a game-changing
evolution of the design process.

In such a complex and fast-moving field there are so many
variables. What is a fact is that the science of supercars never
stops; in some sense it is the one area that is as rapid as the cars
it creates. The science accelerates greater performance; greater
performance motivates the science. For example, in 2017, the
new Porsche 911 GT2 RS lapped the Nürburgring in 6 minutes
47.3 seconds; that's *ten seconds faster* than the acclaimed and
remarkable Porsche 918 Spyder of just four years earlier.

Another case in point: the Bugatti Veyron, which held the
title of the fastest series production road car for over a decade.
When presented with the challenge of accelerating from 0 to
250mph and decelerating back to 0, the Veyron would be taken
to VW's Ehra Lessien testing ground, which has a straight about
5.5 miles long and a series of huge banked curves. The Veyron

would rapidly take the bank leading onto the straight at around 125mph, then as soon as the road straightened up, the driver would floor the Bugatti's fast pedal and rocket up to 250mph, then slam the brakes on and decelerate back to 0mph just before the end of the straight.

Not so the Bugatti Chiron.

The Chiron doesn't even need to use the banking at 125mph. It can begin the run at a standstill at the start of the straight.

Then it can accelerate to 250mph, and back to 0 *in just under 2 miles*.

That means the Chiron can go from 0 to 250mph and back *THREE TIMES* in the same distance a Veyron can do the same feat only ONCE.

That is the relentless, at times bewildering and always utterly fascinating science of supercars.

"The science of supercars never stops; in some sense it the one area that is as rapid as the cars it creates."

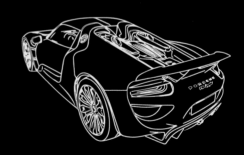

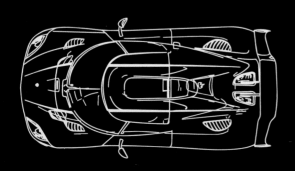

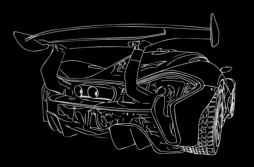

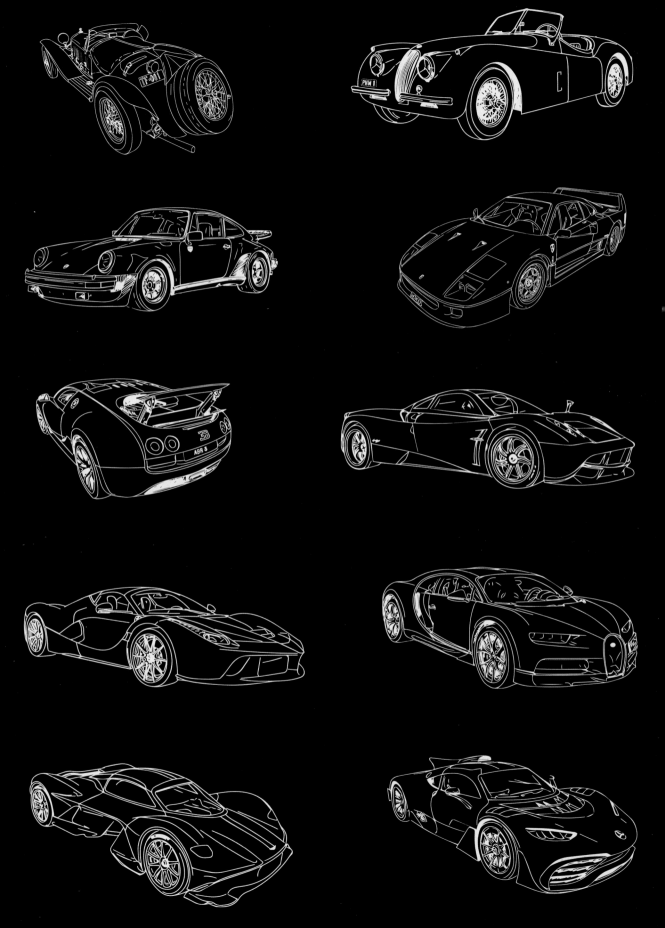

INDEX

PICTURE CREDITS

AUTHORS' ACKNOWLEDGEMENTS

Special thanks to David Coulthard for his Introduction and hugely appreciated, continued support of Martin Roach's work.

The primary research and first-hand material in this book was only made possible by the kind collaboration and immense expertise of all the esteemed interviewees. For this, the authors would like to heartily thank (in alphabetical order):

Achim Anscheidt, Valentino Balboni David Coulthard, Dr Wolfgang Dürheimer, Chris Goodwin, Tony Hatter, Steve "The Legend" Higgins, Liam Howlett, Christian von Koenigsegg, Michael Kodra, Julius Kruta, Nigel Mansell, Professor Gordon Murray, Adrian Newey, Horacio Pagani, Michael Quinn, Peter Read, Rob Rowsell, Dr Kerry Spackman, Andy Wallace, Magnus Walker, Dr Frank Walliser, Nick Wirth, The Royal Automobile Club, The British Racing Drivers' Club

Martin Roach would like to thank (in alphabetical order): Conrad Allum and all at Jaguar Land Rover, Alan Bodfish and the WO Bentley Memorial Foundation, Nicole Carling and all at Red Bull, Mike Burtt, Paul Chadderton, Brian Davies, Rob Durrant and all at Porsche, Jo Hassall, Adrian Davies and all at Bentley Motors Limited, Keith Holland and all at McLaren, Manuela Hoehne, Marie-Louise Fritz and all at Bugatti, Gerhard Heidbrink and all at Mercedes-Benz Classic, Catrin Dunz, Birgit Zaiser and all at Mercedes-AMG, @theghostoutlaw, Stephen Griggs, Ian Hunt, Nic Mira, all at Gordon Murray Design, Mick Pacey and all at Export 56, Christina Lange, Philip Porter, Abigail Humphries and all at Porter Press International (www.porterpress.co.uk), *Jaguar E-Type – The Definitive History*, by Philip Porter, *Ultimate E-type – The Competition Cars* by Philip Porter, and *E-type Jaguar DIY Restoration and Maintenance* by Chris Rooke, all published by Porter Press International, Simon Priest, Pierre-Henri Raphanel, Giulia Roncarati, Hannes Zanon and all at Pagani Automobili SpA, Sandra Da Rosa Pinto, Peter Saywell and all at Saywell International, Lee Sibley and all at Total 911, George and Henry Thomas and all at Lakeside Classics, Steven Wade and all at Koenigsegg, Kevin Watters and all at Aston Martin Lagonda, all at Wirth Research.

A very big thanks to our magnificent publishing team at Octopus who have worked tirelessly on this project, including Joe Cottington, Alex Stetter, Jack Storey, Jeremy Tilston, Allison Gonsalves, Matthew Grindon and Karen Baker.

Many thanks to Carol Neath for all the invaluable behind-the-scenes work and support.

A final special thank you to Trevor Dolby, who acted on instinct by commissioning my Bugatti book idea many years ago, a leap of faith for which I will always be very grateful. With love and respect to George Dolby.

Neil Waterman would like to thank (in alphabetical order): Robin Gearing, Asad Noorani, Francesco Perini, Patricia Potts, Rob Rowsell, Daisy Waterman, Pam Waterman, Will Waterman, Nick and Louise Wirth. I'd also like to reflect Martin and John's thanks to so many amazing, learned and helpful people. With special thanks to David Coulthard for the personal introduction to Martin Roach and the notion that we might one day work on a book together. Good call! Finally, thanks to brother Jules and all at Waterman Graphics.

John Morrison would like to thank (in alphabetical order): Dominic Amian, David Bagley, Valentino Balboni, Nigel Barrett, Jürgen Barth, Sam Beale, Laura Beduz, Derek Bell, Sandi Boyland, British Racing Driver Club, Creighton Brown, Michael Burtt, Tiziano Carugati, Jonathan Carter, David Clark, Mac Daghorn, Brian Davies, Alberto Giovanelli, John Greasley, Gue Family, Anthony-Robert Hatter, Max Heidegger, Steve Higgins, Jane Holmes, Ian Hunt, Barry Jell, Julius Kruta, Nicholas Lancaster, Rainer Merlot, Martin Ladbroke, Peter Lovett, Peter Lyons, Alfreda I Morrison, Derek Ongaro, Franz-Josef Paefgen, Horacio Pagani, Pagani Family, William Earl of Pembroke, Tom Peters, Dr Ferdinand Piëch, Bugatti Prescott, Stuart Pringle, Pierre-Henri Raphanel, Surjit Rai, Natasha Rayn, Peter Read, Brian Redman, Charles, Duke of Richmond and Gordon, Giulia Roncarati, Royal Automobile Club, Julian Sainsbury, Peter Saywell, Dr Wolfgang Schreiber, Jeremy Snowdon, Alasdaire Stewart, Stephanie Sykes-Dugmore, Franco Utzeri, Malcolm Vaughan, Tom Walkinshaw, Ian Williams, Tony Willis.